Portfolios for Technical and Professional Communicators

Herb J. Smith
Kim Haimes-Korn
Southern Polytechnic State University

Contributing author
Betty Oliver
Southern Polytechnic State University

PEARSON

Prentice
Hall

Upper Saddle River, New Jersey
Columbus, Ohio

Library of Congress Cataloging-in-Publication Data

Smith, Herb J.
 Portfolios for technical and professional communicators/co-authored
by Herb J. Smith & Kim Haimes-Korn.—1st ed.
   p. cm.
 Includes bibliographical references and index.
 ISBN 0-13-170458-3
  1. Communication of technical information. 2. Employment portfolios.
I. Haimes-Korn, Kim. II. Title.
T10.5.S635 2006
650.14—dc22

                                                    2006007770

Editor: Gary Bauer
Editorial Assistant: Jacqueline Knapke
Production Editor: Kevin Happell
Design Coordinator: Diane Ernsberger
Cover Designer: Ali Mohrman
Cover art: Corbis
Production Manager: Pat Tonneman
Senior Marketing Manager: Leigh Ann Sims
Senior Marketing Coordinator: Elizabeth Farrell
Marketing Coordinator: Melissa Orsborn

This book was set in 10/12 Meridien Roman by Techbooks, New Delhi, India Campus.

Pearson Education Ltd.   Pearson Education Australia Pty, Limited
Pearson Education Singapore Pte. Ltd.   Pearson Education North Asia Ltd.
Pearson Education Canada, Ltd.   Pearson Educación de Mexico, S.A. de C.V.
Pearson Education—Japan   Pearson Education Malaysia Pte. Ltd.

ISBN 0-13-170458-3

For Larry, Garrett, and Maggie
and my mother, Bobbi

For my parents, Herb and Ellen

Preface

Welcome to *Portfolios for Technical and Professional Communicators*. The purpose of this textbook is to provide technical and professional communication students, as well as practitioners, with a short, practical guide for creating, revising, and using paper and electronic professional portfolios. The portfolio-building process is described in detail, including a variety of student examples to support concepts and guidelines. In addition, this book discusses legal and ethical issues that impact the portfolio-building process, along with information on how to conduct a successful job search and how to use portfolios during the interview.

Special Features of This Book

This book has three distinct features. First, it focuses on a process approach to creating professional portfolios and is written specifically for technical and professional communicators. Second, it has evolved from five years of team teaching a project portfolio capstone course designed specifically for technical and professional communication majors at our university. While taking the course, each student designs and creates both an electronic and a paper portfolio. During the years that we have been teaching this course, we have developed and refined course materials and have introduced new teaching strategies. While writing this book, we tested sample chapters in the classroom, inviting student feedback that we have incorporated in the book. Students' success stories support this class and are included in the text. Also included in the text are a wealth of exercises, examples, and assignments.

The third distinct feature is the perspective that the authors' combined expertise brings to the book. Kim Haimes-Korn coordinates a composition program. Her expertise is in rhetoric and composition, with research interests in portfolio theory and practice. Herb Smith coordinates the undergraduate technical and professional communication degree programs, is an active member of the Society for Technical Communication and the Association of Teachers of Technical Writing, and has technical communication industry experience. Both authors have delivered presentations on portfolio design at national conferences, and have taught and consulted in the technical communication profession collectively for over 20 years. In every way, this book is a collaboration. Each author contributed equally to its contents.

Finally, we thank the many mentors who volunteered their time and professional guidance to the students as they developed their portfolios. Specifically, we thank David Anthony, Susan Barclay, Tom Burns, Jack Butler, Karen Docherty, Verna Hankins, Wanda Kane, Scott McMichael, Catherine Mesa, Michael Perry, Nancy Silver, Lynne Stockton, Mirhonda Studevant, Bobby Vaughn, and Michel Alexander.

Methodology

We consider this book unique to the field because of the inclusion of student voices and portfolio samples. The data described in this book were collected over the five years of teaching the professional capstone course in our degree program. The methodology includes naturalistic inquiry from a teacher-researcher perspective. Rather than a case study or clinical approach, student voices are presented throughout the book as they pertain to particular issues. Our research focused primarily on student work and includes an extensive References and Resources section at the end of the book. Throughout the book, we invite you to get to know our students and use their first names (when given permission) to identify their words and work.

Audiences for This Book

This book will be helpful for students and practitioners who want to develop a paper and/or electronic portfolio to pursue any of the career paths in technical and professional communication such as the ones listed below:

- Professional communication
- Technical communication
- Web development
- Graphic design
- Multimedia development
- Publication management

- Journalism
- Public relations
- Instructional design
- User assistance documentation
- Video production

How to Use This Book

This textbook is organized into chapters that describe the portfolio-building process, including the use of professional portfolios during the job search. Because each chapter contains its own objectives, guidelines, examples, exercises, and assignments, each can be read as a separate unit.

Chapters 1, 2, 3, 4, and 7 focus on the portfolio-building and revising processes. Chapter 6 discusses some of the important intellectual property issues that affect the creation of portfolios. Chapter 5 focuses exclusively on the creation of electronic portfolios, and Chapters 8 and 9 discuss how to use the portfolio during the job search. We have organized the chapters in a linear sequence, but we recognize that readers may prefer to read the chapters in a different order, depending on their specific needs. For example, you may want to review basic career-related information, including tips on how to write or revise your resume, before you begin designing the content of your portfolio. If so, you will want to read Chapter 8 before reading Chapter 3. Chapter 6 can be read at many different points as you create and revise your portfolio materials. Similarly, the chapter on revision (Chapter 4) may be helpful in the process of drafting and finishing your portfolios. The design of the book is fluid enough so that you can sequence the chapters to meet your personal needs. The following annotated table of contents briefly describes each chapter in the book.

Note: Every effort has been made to provide accurate and current Internet information in this book. However, the Internet and information posted on it are constantly changing, and it is inevitable that some of the Internet-related information in this textbook will change over time.

Annotated Table of Contents

Chapter 1: Understanding Professional Portfolios: An Overview

This chapter reviews current theory and practice related to portfolios and works to complicate and form working definitions and possibilities for portfolios in both academic and professional settings. It examines the development of the portfolio as a professional sales kit for technical and professional communicators in the twenty-first century, introducing readers to the processes they will engage in as they develop their portfolios.

Chapter 2: Creating a Portfolio Identity

This chapter guides students through the process of creating an identity and a professional persona for their portfolios. These concepts are explored through heuristics and personality tests that ask them to define their particular niche within the field and to develop their own sense of style and direction. Students will also work to define the rhetorical context for their portfolios as they assess their purposes, audiences, and strategies.

Chapter 3: Portfolio Contents, Design, and Structure

This chapter helps students make decisions about the contents, design, and structure of their portfolios, using the guiding principle that portfolios are representative rather than comprehensive, and helps them choose materials that best represent their skills, goals, and identities. Using samples and a rubric, the chapter introduces ways to construct a proposal that outlines the contents, design, and structure of the portfolios. Students will come to understand the concept of a creating a metaphor for their portfolios to act as a guiding design principle.

Chapter 4: Revising for Portfolio Quality

This chapter helps students to understand the methods they can use in revising their documents and artifacts to achieve portfolio quality. It guides them through rhetorical and design decisions that enables them to show a variety of skills to a new audience. The chapter discusses the importance of both local and global revision strategies, including the creation of context and thematic organization.

Chapter 5: The Electronic Portfolio

This chapter introduces students to electronic portfolios. It analyzes the differences between the paper and electronic versions and includes instruction on Web design, graphic elements, and navigation guidelines. Students will also design a working structure and template for their electronic portfolio.

Chapter 6: Legal and Ethical Issues Affecting Portfolios

This chapter outlines the legal and ethical concerns related to portfolios. Students will come to understand how issues of legality will guide their decisions in their document design and the creation of their portfolios. The chapter covers copyright, image use, and ethical choices.

Chapter 7: Getting Feedback: Responding to and Revising Portfolios

This chapter provides advice on finishing the first draft of the paper and electronic portfolios. Students are given guidelines for an effective response from other students and/or colleagues. The chapter also introduces the concept of professional mentors as responders and addresses ways to set up and nurture mentor relationships. Theories of revision are discussed as well as the ways students can reshape and polish their portfolios for professional presentation.

Chapter 8: Portfolios and the Job Search—Getting Prepared

This chapter provides tips on how both beginning professional communicators and seasoned veterans can use their portfolios during the job search. It discusses how to develop a marketing plan and guides students through the process of creating the documents they will need to conduct an effective job search, including resumes, cover letters, and portfolio formats.

Chapter 9: Using Portfolios During Interviews

This chapter guides students through the interviewing process and comments extensively on how to use portfolios effectively during the interview. It examines how paper and electronic portfolios have replaced the traditional writing sample, always a required document at the interview, with a more extensive and varied collection of technical writing samples that show multiple skill sets. The chapter provides tips for professional communicators on how to present both the paper and electronic portfolios to maximum advantage.

Conclusion

This chapter reviews the purposes and uses of the portfolios and restates the goals of the book. It concludes with the idea that the portfolio is an evolving document that is reshaped over time. Former students and professionals discuss the ways they build and use their portfolios throughout their careers.

References and Additional Resources

Appendix

Acknowledgments

We thank all the students who took the project portfolio course. Without their support and feedback, we could not have written this textbook. In addition, we extend a special thank-you to the students whose voices and projects appear in the book: Miranda Bennett, Tom Burns, Brian Dooley, Angela Dixon, Judith Dickerson, Norma O. Gonzalez, Jarmon Gray, Amy Grau, Wylie Jones, Joy Leake, Nanette Packman, Wilda Parker, Trina Queen, Michel Alexander, Sarah Milligan Weldon, and Brian Wray.

We also acknowledge the contributions of Dr. Betty Oliver, who wrote Chapter 5 on electronic portfolios. Dr. Oliver has assisted us in developing the electronic portfolio portion of the course and has used portfolios in her teaching for many years. We gratefully acknowledge the enthusiastic support of Kenneth Rainey, our department chair; Alan Gabrielli, Dean of Arts and Sciences; and our colleagues in the Humanities and Technical Communication Department at Southern Polytechnic State University, who believe in the value of a portfolio course within our curriculum.

About the Authors

Herb J. Smith, Ph.D., is Associate Professor of Humanities and Technical Communication at Southern Polytechnic State University in Marietta, Georgia, where he is the Coordinator of the Bachelor of Science and Bachelor of Arts degrees in Technical and Professional Communication. He teaches a variety of courses in both the undergraduate and graduate programs, including technical writing, business communication, project portfolio, user documentation, and technical training.

Dr. Smith has delivered professional presentations on portfolios at two national technical communication conferences and has served on the editorial review board for *Technical Communication Quarterly*. He was also the principal author of a successful National Science Foundation grant and was guest professor of technical writing in Germany in 1998. Dr. Smith has published widely in most technical communication journals, and his article "German Academic Programs in Technical Communication" appeared in *Technical Writing and Communication* (Winter 2003).

Kim Haimes-Korn, is Associate Professor of English and Director of the Writing Program at Southern Polytechnic State University. She holds a Ph.D. and an MA from Florida State University with a concentration in Rhetoric and Composition. Dr. Haimes-Korn currently teaches in both the graduate and undergraduate Humanities and Technical Communication degree programs. She teaches courses such as Project Portfolio, Rhetorical Theory and Practice, Composition, Writer's Workshop, and Small Group Communication. She has also developed courses for the computer writing classroom in which students creatively integrate the processes of writing with technology. Dr. Haimes-Korn's teaching philosophy encourages dynamic learning and focuses on students' power to create their own knowledge through language.

Much of Dr. Haimes-Korn's scholarship in the field focuses on classroom inquiry and draws upon student voices to understand writing theory and practices. She writes on response theory, composition pedagogy, collaborative learning, portfolios, visual literacies, and multicultural pedagogies. Her scholarship on portfolios includes presentations at national conferences along with many teacher-training workshops on teacher and student portfolios. Some of these titles include "Teachers Talk: Portfolio Perceptions and Practice at a Four-Year University," "Teacher Portfolios as Sites for Reflection and Assessment," "Creating Student Portfolios: Theory and Practice," and "Student Portfolios Across the Disciplines."

Contents

1 Understanding Professional Portfolios: An Overview

The portfolio is a culminating exercise that not only demonstrates every single skill a student learns in her program and at the caliber at which she performs but is also useful to the student once the class has ended.   Michel

I was very surprised at the amount of creativity that went into developing the portfolios. They aren't just collections of various projects; they're a reflection of the people who created them.   Brian D.

I have learned that a portfolio is a unique expression and a collection of someone's work.   Nanette

INTRODUCTION

Like the students just quoted, you will come to understand your professional identity and how to create a distinctive professional portfolio. Before you create your own portfolio, it is important to place the idea of portfolios in a larger context. To provide this context, Chapter 1 covers the following topics:

* What is a portfolio?
* What do all portfolios have in common?
* Why consider portfolios?
* Portfolios and career professionals
* Portfolios and technical and professional communicators
* Types of portfolios
* Portfolio formats
* Organizational strategies
* Analyzing portfolios

The information in Chapter 1 describes portfolios in terms of what they are and how they can function in your professional life.

WHAT IS A PORTFOLIO?

In recent years the term "portfolio" has been used in many different contexts and situations. Educators have recognized the value of evaluating students in terms of their overall performance and their ability to articulate and demonstrate their skills through portfolios. Professionals use portfolios for purposes such as job searching, performance evaluation, and client presentation. As a job-search tool, a portfolio enables you to go beyond your resume and demonstrate what you can do as well as what you know. Portfolios play an important role in many fields and disciplines, from the arts and architecture to technical and professional writing. Although they have existed for several years in educational settings, portfolios are now entering the mainstream as integral tools within the educational and professional worlds. The role that portfolios play in education is constantly changing as new ideas and uses alter their form and function.

Before you begin to create your own portfolio, it is important to start with a working definition of the term. A portfolio can be defined as a distinct, representative collection of work for a particular audience and purpose. It is a representative, not comprehensive, collection of your best work, an evolving collection of artifacts that will change as you change. A portfolio mirrors your professional growth and development, and you should maintain it to showcase the skills and talents that best fit the position you are applying for. A well-designed portfolio will address the following points:

* What you did
* Why you did it
* How you did it

Peter Elbow and Pat Belanoff, well-known advocates for portfolio use, note that portfolios have value because they present different writing samples created under different conditions, thus providing a more trustworthy picture of the individual's writing ability than a single writing sample (Yancey and Weiser 1997, 25–26). Kimball defines a portfolio as a "reflective collection of work" having a specific purpose and used to prompt feedback (Kimball 2003, 5). Wendy Bishop notes that portfolios are important because they help writers identify a "variety of audience needs" and develop maturity of expression (Bishop 1991, 25). Commenting on the valuable role that portfolios play during an interview, a marketing director of a software company observes that portfolios are very important, particularly for newly minted professional communicators, because they are the only way to prove that you can deliver on the promises made during the interview.

Whether you are building your portfolio from scratch or revising an existing portfolio to keep it current, make sure that your portfolio identifies who you are and what area of technical or professional communication you may want to pursue, whether it is a career in multimedia, journalism, web design, software documentation, medical writing, or some other area.

WHAT DO ALL PORTFOLIOS HAVE IN COMMON?

Many educational theorists support the use of portfolios throughout education and in the job market. The following statements reflect general assumptions that these theorists and practitioners have in common portfolios . . .

* Engage writers/professionals in active reflection and articulation of their development and learning. This reflection might take the form of a process statement, a letter of self-evaluation for a school setting, or an introduction to a professional portfolio.
* Contain different types of work (writing, graphics, projects) that showcase a variety of skills (formal and informal) of a particular writer, designer, or group of writers.
* Create opportunities for writers to take ownership through representative selection and presentation.
* Reflect clearly articulated goals of a particular community, classroom, program, profession, or role.
* Reflect the distinct individualism of a body of work.
* Evolve as the writer/professional develops and gains experience.

WHY CONSIDER PORTFOLIOS?

Here are some of the reasons to consider portfolios:

* Portfolios create a portrait of a professional writer within a particular context such as a class, program, profession, or organization.
* Portfolios encourage professional writers to shape their writing for multiple purposes and audiences.
* Portfolios give professional writers choices in how to present themselves and, in turn, provide a stronger sense of ownership of their work.
* Portfolios help professional writers to articulate their learning through reflective practices and statements.

- Portfolios present writing as both a process and a product.
- Portfolios reveal development over a period of time.
- Portfolios allow students, teachers, and employers to concentrate on overall writing abilities/issues/skills rather than just evaluation—allowing more room for innovation and experimentation.
- Portfolios encourage professional writers to bridge concepts, disciplines, and fields.

PORTFOLIOS AND CAREER PROFESSIONALS

Portfolios have been used for roughly 20 years in some professional fields because they are an important career tool. This section comments briefly on how the following professions use portfolios:

- Architects
- Graphic designers
- Teachers
- Engineers

Architects

Architects often create design portfolios that focus on architectural renderings, landscape designs, or environmental designs. These design portfolios can be used in a variety of situations, including employment interviews, grant applications, or applications to graduate school. As Linton notes, graduate schools of design often require portfolios as part of their admission packets (Linton 2003, 14).

Graphic designers

Graphic designers have used portfolios for many years. A reputable institute in Atlanta, for example, requires that its graphic design majors, video production majors, and digital media production majors create portfolios to show potential employers.

Teachers

Teachers are often required to create assessment portfolios for promotion and tenure, and student teachers often create portfolios to use during their job search. Teaching portfolios may include the following documents:

- A statement of the candidate's teaching philosophy and goals
- Lesson plans and handouts
- Graded student work
- Videotapes of classroom teaching
- Reflective narratives on teaching performance

Engineers

While the primary audiences for this book are technical and professional communicators or those pursuing a minor or concentration in either of these two areas, you may be an engineering major who understands the important role that effective communication skills will play in your career. Portfolios are now being used as assessment tools in engineering programs throughout the country, and this practice is likely to continue.

As Scott and Plumb note in "Using Portfolios to Evaluate Service Courses as Part of an Engineering Writing Program," the Accreditation Board for Engineering and Technology (ABET) 2000 criteria require engineering programs to document that their graduates have the ability to communicate effectively (Scott and Plumb 1999, 337–38). For example, the University of Washington's College of Engineering has established an extensive portfolio assessment project, described in detail by Scott and Plumb, to measure student learning throughout the curriculum. The project is called the Portfolio Evaluation Project (PEP). PEP's goals include collecting information that sheds light on the writing experience of College of Engineering students and creating performance outcomes for engineering writing at the University of Washington ("ABET 2000 Information," University of Washington College of Engineering Web site).

If you are in an engineering field, you may want your portfolio to focus on your engineering projects and design work, perhaps including a tools page that shows your skills using MicroStation, Mathcad, or Excel. You may also want to devote a section of your portfolio to your communication skills because good communication skills are an important factor in promotions. This book provides tips on how to build a portfolio that documents your effective communication skills in engineering reports and other workplace documents. And whatever your professional field, this book is designed to provide tips on portfolio building that you can modify to fit your own professional goals.

PORTFOLIOS AND TECHNICAL AND PROFESSIONAL COMMUNICATORS

As noted earlier, portfolios are not new. However, they are relatively new to the technical and professional communication fields.

A decade ago, job descriptions for professional communicators emphasized excellent writing and editing skills supplemented by the ability to present technical information in a format readily understood by nontechnical users. To demonstrate these skills, technical and professional communicators brought some writing samples, often collected in a cluttered briefcase or tattered manila folder, to the interview. Because such samples cannot present a clear picture of your talents and skill sets, they cannot market your strengths effectively at an interview. What can an interviewer learn about you from flipping through a 300-page user guide? What interviewer wants to look at a 100-page formal proposal? No matter how effective you may be at organizing your writing and design samples, it will be difficult to present a clear picture of who you are and what you can do without a professional portfolio.

For professional communicators in the twenty-first century, professional paper and electronic portfolios are more important than ever before, given the wide range of career choices that technical and professional communicators have. "Technical and professional communication" is an umbrella term that encompasses a number of fields. These fields include the following:

- Corporate training
- Graphic design
- Information design
- Instructional design
- Journalism
- Multimedia development
- Proposal writing
- Publication management
- Public relations
- Technical editing
- Technical writing
- Usability testing
- User support documentation
- Video production
- Web development

This list is not all-inclusive; as products and technology change, so will the field of technical and professional communication. However, the list will help you better identify the specific field or fields. Once you have identified these fields, you will be better able to select the types of artifacts that you may want to include in your portfolio. While it is a good idea to make sure that your portfolio demonstrates versatility and a range of skills and tools, you may want to devote a section or two of your portfolio to one, two, or three of the fields noted previously. If you have experience with different types of computer programs, such as RoboHelp, Dreamweaver, PageMaker, Flash, FrameMaker, or Illustrator, include in your portfolio artifacts that show your knowledge of them. Table 1.1 provides examples of documents that might appear in some of the technical and professional communication fields.

TABLE 1.1
Technical and Professional Communication Field and Possible Portfolio Documents

Field of Technical and Professional Communication	Possible Portfolio Documents
Technical Writer and Technical Editor	A range of documents including marketing pieces, research reports, memos, proposals, and user documentation.
Graphic Designer Web Developer	A range of stand-alone graphics (e.g., photos, Web pages, line drawings) created with tools like Photoshop, Illustrator, Flash, and Dreamweaver. Also include documents using spot color and graphic features—fonts, background graphics, and color.
Multimedia Developer Video Producer	A CD or URL with sample multimedia projects supporting different project goals (e.g., training, information sharing, marketing). A storyboard or storyboards showing project design. Emphasize sound, text, graphics, and motion working together. Include streaming video clips.
Publications Manager	A variety of print and electronic documents showing how you can design messages for different audiences. Pieces might include press releases, a homepage for a hypothetical company, business letters, brochures, logos, and business cards.
Instructional Design	For instructional design, you might focus on an interactive training tutorial, a set of instructional objectives, assessment materials, or sample exercises. You may want to include a script for a training manual, a video of a training session you taught, or a PowerPoint presentation used during a training session.

TYPES OF PORTFOLIOS

Based on their content and use, portfolios can be classified into four types:

* Working portfolios
* Academic portfolios
* Assessment portfolios
* Professional portfolios

Working portfolio

A working portfolio is exactly that—a portfolio in its roughest form, an informal loose collection of artifacts that you have created for various reasons. It is formed during your writing course, degree program, or career. This portfolio might include working drafts, invention exercises, and process statements. Because a working portfolio is a repository for a wide range of documents, it is often a large collection of work that is not organized according to any set criteria or genre. It may contain, among other things, poems, drawings, stories, reports, multimedia pieces, and graphics. Often these pieces will need significant revision. The primary audience for a working portfolio is you, the author of the documents, and perhaps a teacher or mentor who may help you select and group documents according to some preliminary plan.

A working portfolio holds the raw material for the other types of portfolios. If you are just beginning your academic studies in technical or professional communication, you should start creating your working portfolio early and add to it something from each course in your major. You may want to create an electronic folder called a "working portfolio" and use it to collect your documents. Try to keep all paper and electronic artifacts so that you will have an extensive resource to draw from when you create an academic and/or professional portfolio at a later date. Always create several backup copies of electronic artifacts to protect against loss or corrupted files.

Academic portfolio

An academic portfolio is a collection of your work for one or more courses and demonstrates your growth and development in relation to a particular subject. It is generally used for response and evaluation purposes in the classroom. If you took a foundations of graphics course where you created several graphics using different software, such as Adobe Photoshop or Adobe Illustrator, you may want to save these artifacts in an academic portfolio labeled "Foundations of Graphics." An academic portfolio may contain suggested revision from your teachers and perhaps even your peers. You may be required to create an academic portfolio for a senior capstone course such as a project portfolio course. A portfolio created for a senior capstone course may be a retrospective collection of work you've created throughout your academic life showing your growth and development as a professional communicator.

Your academic portfolio may also include artifacts that you have created during an internship, co-op job, or other jobs that you may have had while pursuing your degree. If the work, such as a chapter from a software manual or a brochure, is content sensitive or contains proprietary information, you should get permission to include it in your academic portfolio. Intellectual property issues are discussed at length in Chapter 6. Because more and more academic institutions are now using portfolios as one way of measuring student achievement, there are a number of portfolio sites that can be viewed from university and college Web sites. You may want to search several different websites to view a variety of academic portfolios. Appendix A lists some sample sites.

Assessment portfolio

Assessment portfolios are common in academic settings and are used by students and teachers alike. They represent your efforts to shape your work for particular audiences and purposes, allowing you to demonstrate and articulate learning and performance by showcasing your work. This might be a selection of your best, revised works or an overview of your total work in a class or program. These portfolios are generally used for evaluation (both formative and summative) or to assess work on a particular project or during a specific time period. A student assessment portfolio is a systematic collection of student work measured against predetermined scoring criteria such as rubrics, checklists, or rating scales (Gomez 2000, 1). Representative work in this type of portfolio might include writing samples, solutions to math problems to show problem-solving skills, lab reports to demonstrate an understanding of engineering principles, and social science research reports to demonstrate an awareness of different research methods.

A teacher assessment portfolio, like a student assessment portfolio, contains a wide range of documents demonstrating that the abilities, experiences, and evidence of teaching excellence. These documents might include a statement of teaching goals, lesson plans, handouts, graded student work, and evidence of professional development. Public school or university administrations may use this portfolio to measure the performance of a teacher who is undergoing annual or tenure review or being considered for promotion.

Professional portfolio

The statement "You don't get a second chance to make a good first impression" is probably familiar to you. A professional portfolio will help you make a good first impression because it is your best work in finished form. This portfolio represents your work in a professional field. It should include primarily your strongest finished products and reveal your strengths as a professional in your chosen field. It should also reflect your professional identity and goals. This portfolio can be used in the job market, for hiring purposes, performance evaluation, or as a demonstration of your work in a particular program, project, or time frame. Professional portfolios evolve over time as you gain different work experiences. Recent, more professional work often takes the place of earlier student or entry-level work to reflect changes in your professional life.

Commenting on the impact that his professional portfolio had at an interview, Brian D., a recent graduate of a technical and professional communication program, noted: "A good resume and cover letter will get you in the door. A good portfolio can set you apart from other candidates and will get you the job." You will want to include the work that you are proudest of, but remember that your professional portfolio, as noted earlier in this chapter and discussed in detail in Chapter 3, should be a representative selection of your best work rather than a comprehensive collection of all the documents you have ever designed or created. While the number of pieces that you should include in a professional portfolio is not fixed, you will want to document a range of skills. You may decide to include as many as 10 to 15 pieces of your best work, particularly if you have little or no related job experience on your resume and are interviewing for your first job as a professional communicator. One of the primary functions of a professional portfolio is to validate the information on your resume, since many interviewers today are skeptical about the claims made on resumes. A well-designed professional portfolio can help answer questions that a resume may raise by clearly identifying who you are, what you have done, and what you can do for a particular employer.

Professional portfolios are also widely used for performance reviews, in promotion packages, and, if you have a consulting business, in marketing your business to prospective clients. Table 1.2 summarizes the distinguishing features of each of these portfolios. While commenting on a recent promotion to technical writer, Amy notes that "part of what got me the job was showing them the portfolio I had created at school and had

TABLE 1.2
Types of Portfolios

Type	Description
Working Portfolio	A loose collection of artifacts consisting mainly of unrevised work. It lacks any clear organization and contains the raw materials for other types of portfolios. The primary audiences are the writer and perhaps a teacher or mentor.
Academic Portfolio	A collection of artifacts often grouped by course content or genre. This portfolio contains selected pieces from a working portfolio designed to show the writer's development as a technical and professional communicator. Selected pieces often include documents with teachers' and peers' comments. The primary audiences are the writer and his or her teachers, parents, and friends.
Assessment Portfolio	A representative collection of pieces used to formally evaluate a student's performance according to specified learning outcomes or a teacher's effectiveness according to predetermined criteria.
Professional Portfolio	A collection of the writer's best work in final form that can be viewed in one or more formats (paper, CD, or Web-hosted). Work should be representative pieces rather than a comprehensive collection that serve as examples of skills, tools, and experience claimed on the resume. The primary audiences are the interviewer, potential client, or supervisor.

working at my old position within the company." In your professional portfolio, you may even want to include a brochure that markets you as a professional communicator.

Your professional portfolio should have a welcome page or introduction that explains the purpose of the portfolio, tips on how to view it (if it is in electronic form), and any unifying design feature such as a metaphor or theme that you have used as a nonprint unifying principle for the artifacts in the portfolio. Various kinds of metaphors and tips on how to use them effectively in portfolios are discussed at length in Chapter 3. Table 1.2 summarizes the distinguishing features of each of these portfolios.

In addition, there are thousands of Web sites that you can visit to view different types of professional portfolios or get tips on how to create your own. To view a wide range of portfolios within the field of professional communication, you should develop a list of search terms to use, perhaps through a search engine like KartOO, to view professional portfolios in the fields included under the umbrella term "technical and professional communication." Sample search terms include the following:

- Portfolios and professional communication
- Portfolios and student samples
- Portfolios and multimedia skills
- Digital portfolios and advertising

PORTFOLIO FORMATS

Paper portfolio

While the digital portfolio may be your preferred format, it would be wise to have your portfolio available in both a digital and a paper format. A paper portfolio is very useful when you are interviewing in person, meeting a client, or preparing for a promotion. This portfolio is one that you carry with you. Generally, work is placed in an oversized notebook or leather-bound, zippered case. Figure 1.1 shows Joy's portfolio in its zippered case. Artifacts are usually displayed on large pieces of paper protected by sheet covers. In many cases, they consist of original documents and can provide a visual and tactile presence. It is best to use this type of portfolio during interviews, presentations (for consultants), and performance reviews. Figure 1.2 shows sample pages from a paper portfolio.

A paper portfolio, generally housed in a ringed binder, is easy to tailor for the specific job requirements of the employer. You may want to include multiple copies of your resume so that you can leave a copy behind

FIGURE 1.1
Sample Portfolio in a Zippered Case
Source: Used with permission from Joy Leake.

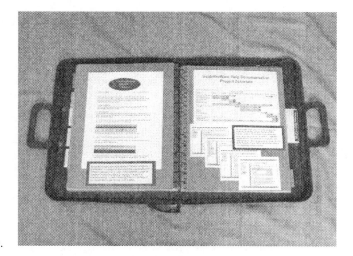

FIGURE 1.2
Sample Pages from a Paper Portfolio
Source: Used with permission from Joy Leake.

after the interview. While your paper portfolio should be representative rather than comprehensive, your versatility should be emphasized by showing your full range of skills. For long documents like manuals and formal proposals, include a complete table of contents supported by a few representative pages from the document. If the job calls for a particular document skill, you may want to include a longer segment of the manual or proposal as an attachment.

Advantages of the paper portfolio.   A major advantage of the paper or hard copy portfolio is that it allows you to structure the interview even if the interviewer hasn't asked you to bring a portfolio. Paper portfolios work best in face-to-face settings because they provide talking points and the ability to focus the discussion in ways that are not necessarily linear. For example, they allow you to skip around and emphasize one skill over another as the conversation grows. They also provide more control during interviews as you come to the table ready to talk about and show your work. Chapter 9 discusses in detail how to present the portfolio during the interview. You may want to include a brief summary statement for each piece that notes why that piece was selected for the portfolio, the audience for it, and any skills or tools used to create the piece that clearly match the job requirements. Some candidates also include a compact disc version of their portfolios in a cellophane (acetate) pocket at the front or back of the paper portfolio.

Electronic or digital portfolio (e-portfolios)

More and more professional communicators are presenting their portfolios in a digital format. The electronic or digital portfolio, often posted on the Internet as a Web page or stored on a disk, is essential in our electronic world. Even in the writing-heavy professions, employers are looking for candidates who demonstrate state-of-the-art skills and basic computer literacy. One major advantage of the digital portfolio is that the format itself clearly demonstrates the high-end software tools or skills listed on the resume. In addition, a digital portfolio may be the only vehicle that can effectively display Web design and other electronic communication skills such as online help systems, clips from streaming video projects, Web sites, or even Web-based training packages.

Like the paper portfolio, the electronic version showcases your work. Unlike the paper portfolio, it will generally be viewed when you are not around. This type of portfolio is very useful because it is convenient. It can also act as a reminder or follow-up after a face-to-face meeting. It gives your audience a chance to ruminate, look more closely (and actually read documents—should they desire), and view at their own pace. It is also an inexpensive way to store your work, as it is quite costly to send out actual samples to prospective employers or clients. (See Chapter 5 for an in-depth discussion of electronic portfolios.) There are two major formats for electronic portfolios:

* Compact disc (CD) portfolios
* Web-hosted portfolios

CD portfolio

CDs are becoming popular as a format for presenting electronic portfolios. The CD portfolio is inexpensive and easy to maintain. You can leave it behind at an interview, knowing that the cost of making another copy is minimal. CD portfolios are also easier to maintain than paper portfolios, and they allow you to display larger documents, like manuals or proposals, without adding bulk. These portfolios are easy to send and are now available in a business-card size so that you can send them or leave them behind with your business card. CD portfolios, however, must stand alone; you will not be there to present their contents. As a result, your CD portfolio probably should have more of a marketing tone or slant than your paper portfolio. Also, a potential employer may want you to send a CD portfolio in order to review your qualifications before committing to the expense of an in-person interview.

Web-hosted portfolio

Almost all the tips for CD portfolios apply to Web-hosted portfolios. Web-hosted portfolios also have some distinct advantages and disadvantages that other portfolio formats don't have.

Advantages. A Web-hosted portfolio provides the widest possible audience for marketing your qualifications and skills. Daily, millions of people surf the Internet, and some will click on your Web site. One or more of these visitors may be potential employers who, after viewing your online resume and portfolio, may invite you in for an interview. This happened to Brian D., who was contacted for an interview by an employer based exclusively on his online resume and portfolio. Brian was offered a technical writing position because he had his resume and portfolio on his Web site; the power of the Web could work to your advantage as well. Having a Web-hosted portfolio also allows you to place the Web URL on your resume, application letter, and business card.

Disadvantages. Of course, there is a downside. You may not want everyone to have access to your portfolio. If this is the case, you can password protect your portfolio. One of the biggest potential drawbacks to a Web-hosted portfolio may be the technology itself. A potential employer may not be able to access your portfolio when needed. As Steven Kendus notes in "Developing a Web-Based Portfolio," it is a good idea to provide the employer with your phone number or e-mail address should technical problems occur (Kendus 2002, 7). You may also want to check to see if the Web site hosting your portfolio is free from advertisements and cookies that might irritate your viewers. Once your portfolio is hosted on a Web site, test it to make sure that your links work. You may also want to use file types like pdf, doc, and ppt that will be familiar to your user.

ORGANIZATIONAL STRATEGIES

One of the decisions you must make about your portfolio is how to organize it. Once you look at your artifacts (the items in your portfolio) and content, you will have a better idea of how you want to arrange portfolio. Remember that your portfolio is more than just a container. It is a whole body of work linked together—a reflection of you as an individual. It will most likely have sections that demonstrate particular skills or ideas. The following organizational strategies might help you begin to shape your work (see the detailed discussion in Chapter 3):

Chronological order

You might set up your portfolio chronologically, from oldest to newest. This structure works well if you want to emphasize progression over time.

Genre

For this structure, divide your portfolio into sections for different genres (or skills)—for example, report writing, creative writing, graphics, and video. Sections can stand on their own or can be viewed as a whole collection.

Subject area

There might be particular subjects that you focused on in your studies or with your clients. Your portfolio sections could represent these different subject emphases. For example, one of our students chose a medical writing section and another decided to emphasize her talents as an investigative reporter. Another dedicated a section to event planning and marketing. Still another wanted to emphasize her fund-raising abilities and dedicated a section to these efforts. This is also a good way to represent a minor or a secondary interest such as management or computer science.

Practice to final product

In the fields of writing and design, employers are looking for individuals with a keen eye on revision. Many jobs call for editing and revising skills and the ability to see a project through from beginning to end. Even if your whole portfolio is not arranged this way, you might consider at least a section that demonstrates these skills. For example, you might include a document before you revised it, complete with editing marks. Another possibility is to show a storyboard for a video or film or even an early version or versions of a logo or Web site design. Narratives of changes and shifts you consciously made as a writer or designer would also be of interest.

Rank order

You can always arrange your work in rank order. Do not arrange it from worst to best, as you might not have time to review all the selections during an interview. Like the chronological arrangement, this structure is best for demonstrating progress over time.

Individual and team projects

Depending on your field and areas of concentration, you might choose to organize your work according to individual and team projects.

Theme/metaphor

As you discover more about yourself and your professional identity (see Chapter 2), you might come up with a metaphor or theme that acts as a unifying design feature for your portfolios. The theme can then be used to divide the sections, which can then include multiple genres and organizational styles within each thematic unit (see Chapter 3 for a detailed discussion on theme/metaphor).

Specialty concentrations

Some people have a stand-alone (or insertable) portfolio that covers a specialty area. This structure works best for people with highly specialized skills. Some of our students created two separate portfolios: one that was more general and one focused on a particular skill set or objective. You can always, of course, rearrange the documents in a single portfolio to call attention to a specialty that you want to emphasize.

TABLE 1.3
Summary of Organizational Strategies

Strategy	Description and Benefit(s)
Chronological Order	Portfolio pieces are arranged from oldest to newest. *Benefit:* Emphasizes development as a professional communicator over time
Genre	Portfolio pieces are grouped by type or category (e.g., report writing, creative writing, graphics, Web documents). *Benefit:* Stresses a collection of related pieces that can be viewed as a whole
Subject Area	Portfolio pieces are grouped by subject area (e.g., medical writing pieces, user support documents, fund-raising pieces, or pieces pertaining to a subject concentration or minor). *Benefit:* Pieces that have a related application are placed together so that they can be reviewed as a whole
Practice to Final Product	Various designs, drafts, and storyboards are sequenced to show how a document evolves from concept to final form. *Benefit:* Emphasizes project development, editing, and revising skills
Rank Order	Portfolio pieces are arranged according to quality. Strongest pieces are placed first, including any work that was published or won an award. *Benefit:* Showcases the best work early in the portfolio
Individual and Team Projects	Portfolio pieces are organized as either team or individual projects. *Benefit:* Emphasizes the ability to work well on a team and the writer's individual strengths
Theme/Metaphor	Portfolio pieces are linked by theme or metaphor. *Benefit:* Provides a consistent look and feel to what otherwise might be viewed as a collection of unrelated pieces
Specialty Concentrations	Portfolio pieces are organized into different portfolios or portfolio sections to emphasize specialized skill sets. *Benefit:* Allows the writer to develop a general portfolio and a more specialized portfolio or portfolio insert to be used as needed
Hypertext Links	Portfolio pieces are presented as a multidimensional structure with links to documents both within and outside the portfolio. *Benefit:* Showcases high-end tools or a different set of design elements and permits display of big documents without adding bulk
Combination	Portfolio pieces are arranged using more than one organizational strategy in order to best display the work. *Benefit:* Creates freedom to use a variety of organizational approaches to showcase abilities and skills

Hypertext links (electronic)

Obviously, your electronic version will require a multi-dimensional structure with links throughout the portfolio and to appropriate outside sources. You will also need to think about site maps, navigation systems, graphics, and templates (see Chapter 5).

Combination

The preceding list provides a way for you to start thinking about the overall structure of your portfolios. This list is not intended to be all-inclusive or discrete. Any of these suggestions might be combined or overlapped throughout your portfolios. You do not have to choose one and stick to it. For example, you might organize the whole portfolio around a theme and then have a chronological or genre structure embedded in the sections. You might also include your specialty areas within this same thematic approach. This list is designed to get you thinking about the possibilities. Table 1.3 summarizes the key features and benefits of each organizational strategy.

ANALYZING PORTFOLIOS

As with all writing and composing processes, there is no one right way to approach the task of creating your portfolios. We present the ideas in this book as guidelines and hope that you will find your own way to structure your portfolios—a reflection of your individual professional identity. As you will see, our students have found many interesting, productive ways to interpret this task; we are happy that they agreed to share their perspectives and experiences. It is through these varied experiences and products that we have formed our own working definition of portfolios. We can introduce you to the theory and methods of design and creation, but it is up to you to apply them to your own work and vision. It always helps to ground theory in reality. We asked our students to familiarize themselves with different kinds of portfolios along with strategies in designing portfolios. They completed this task through an analytical exercise in which they had to find and reflect on online portfolios. We asked them to look for patterns, themes, and approaches, as well as to consider the rhetorical elements of the portfolios and explore issues such as style, editing, organization, audience, and voice. We also asked them to consider the technological and visual dimensions as they analyzed graphics, color, and visual appeal.

We have included some of their responses to this assignment to show you the ways in which they read and compared the samples they chose to analyze. It is important that you conduct this kind of search in order to understand what is happening in the field at the moment because technical and professional communication is a dynamic profession where things are always changing and technology is quickly outdated. Notice the ways in which these students begin to shape their ideas and identify the strengths and weaknesses of the portfolios they reviewed.

We start with a sequence of comments from Miranda to show the ways in which this student built upon her ideas as she searched. We follow this sequence with a more random list of tips and observations from other students as they analyzed the sites. Miranda's initial comments on a technical writer's electronic portfolio that she viewed range from design to organization to graphics:

> I like the clean look of the portfolio, particularly the thumbnails. I also like that he included references to
> pieces that he could not show on the Internet, with an explanation that he could show the "classified" piece in
> person. That intrigues me. I like the bold choice of colors. The blue background isn't too overpowering.

She notices other issues and design elements as she reads several samples and begins to compare them, such as thumbnail explanations and layout:

> The picture of him in the banner is a little large and scary. He's a good-looking guy, but the cutoff at the neck
> looks awkward and his skin tone seems too red. I also think there is too much space on the page between the
> thumbs and their explanations.

Her analysis brings up small issues like the use of a personal picture on an electronic site. Notice the ways that Miranda creates questions for herself as she begins to identify what she likes and dislikes about each portfolio. This early analysis is very open—she is just absorbing and processing what she sees and thinks. Through her search she realizes that she likes a "sleek portfolio" with "soothing" colors. She also liked the subtle way one author chose to include a French version of her resume to "showcase the fact that she's bilingual as well, without being too pretentious." In addition, she noted that this author's e-mail address appeared on every page—a simple but effective strategy she decides to incorporate into her own portfolio.

You can see Miranda responding to design elements, as she records her overall impression of the sites:

> Light gray for some fonts is too light on the white background. The text size doesn't change with the browser. With such small text and such a light color, I think that making it a static size was a bad choice. . . . Wow . . . that blue is really bright . . . especially that much of it. This site is supposed to be for an individual who is marketing himself as a group. It is really a pretty layout, but the colors are just too loud against the black background. It doesn't encourage me to read the pieces he's showing off.

As she gathers data, she moves toward an assessment of her favorite portfolio site in her search:

> *Barking Dog* is by far my favorite in the list of portfolio sites I've looked at. I like how the branding stands out right at the onset. And though a barking dog has nothing to do with technical writing, they make that part clear in the heading. Their portfolio pieces look great. The thumbs are big enough to show them off. The "Do Not Press" shows a quirky sense of humor that is charming but not too over the top. Their use of Flash is nice and not overwhelming. And their spiel is fun to read.

Miranda is beginning to define a structure or a rubric on the design and concept of her portfolio. She is beginning to locate herself within her analysis as she defines her likes and dislikes. She is forming her goals for her own portfolio by analyzing others.

Other students had similar reactions to their searches. The first two student responses focus on search strategies.

> I found all types of portfolios ranging from elementary classes through college classes. To make my search more specific, I began it under various disciplines such as graphic art, Web page design, technical communication, and so on.   *Nanette*

> I learned a lesson in net surfing with this exercise. The Web page gave no option for printing these samples. When they were double clicked, a larger version would be displayed but no button or menu option was provided for printing.   *Tom*

The next four students commented on organizational strategies:

> I really liked it when the person's portfolio explained all the way through what the piece was for and how the piece was created.   *Sarah*

> This portfolio is confusing because you have to read all through her theory passages to link to the examples. If I were an employer, I wouldn't care in the least if she understood theory—I just want to know what her writing's like. I want to see lists and short explanations.   *Michel*

> It is pretty easy to tell which portfolios are successful and which ones are not by looking at the home page.   *Brian D.*

> The simpler ones were the easiest to navigate, therefore the easiest to gain full knowledge of the person's talent.   *Sarah*

Finally, Nanette and Tom also had comments about the portfolio's graphic design features:

> The teacher's portfolio was different from the artist's portfolio because she was displaying more text assignments instead of paintings.   *Nanette*

> Unfortunately, the images in each student's portfolio are tiny JPEGs, and it is impossible to read most of each ad. One can get a fair impression of the quality of a piece by reading the headline and viewing the graphic.   *Tom*

Color and repetition played an important part in the gaining of attention. Almost all of the portfolios I looked at included a picture and a short description of themselves.   *Nanette*

Taken together, these comments illustrate how these students began to make choices about their own portfolios. This analysis shows that the portfolio, like any other form of communication, is dependent upon many dimensions to make it work. There is also no distinct model for imitation. Instead, you will find a full range of ideas, strategies, and methods of presentation. The goal of creating your portfolio will be to find the right elements to express your individual knowledge and skills.

SUMMARY

Many professionals in a variety of fields are discovering the valuable role that portfolios can play in their careers. If you are studying to become a professional communicator, in particular, you will want to create a portfolio that you can update and maintain throughout your professional life. The documents included in your portfolio will help you demonstrate to a wide range of audiences the skills and qualifications you claim on your resume.

Chapter 1 describes portfolios as evolving documents and notes that you may design more than one type of portfolio during your professional life. The tips provided in this chapter should help you accomplish this goal no matter what format or formats you choose for presenting your finished portfolio.

The remaining chapters in this book take you through the process of defining your professional identity and goals, as well as providing a detailed process to help you create your portfolios.

ASSIGNMENTS

Assignment 1: *Analyzing Portfolio Samples*

Conduct and record an Internet search to find sample portfolios and advice on designing, creating, and marketing portfolios. Use different search engines and search terms to conduct a thorough and specific search. As you search, keep a list of what you notice. Focus on elements such as design, writing, organization, navigation, and graphics, along with other criteria gathered from the student comments in this chapter. Create an analysis rubric that shows how your observations fit with your working criteria. Record all URL addresses as you go for future reference.

Assignment 2: *Writing an Annotated Review*

For the Web sites you collected in Assignment 1, write an annotated review or short narrative in which you describe your impressions of these sites and the ways they begin to address your emerging ideas about portfolios. Look critically at the sites and analyze them for strengths and weaknesses. Compare them and begin examining your own ideas about communication, technology, and design. Analyze the ways your impressions fit with your knowledge and beliefs about what constitutes effective writing and graphics. Consider your general impressions along with the conventions and trends within your own discipline. Work to address the criteria you have established through these exercises.

Assignment 3: *Critiquing a Home Page*

This assignment draws upon the work you completed in Assignments 1 and 2. Review your notes and annotated bibliography and choose two portfolio home pages, one that you consider effective and one that you consider ineffective. Print and annotate the home page. In a one-page analysis, describe your impressions as you critique the site. Consider both its strengths and weaknesses in relation to issues as design, content, graphics, color, and navigation, along with your overall impressions of the site.

Assignment 4: *Analyzing Types of Portfolios*

Based on this chapter's content, what type of portfolio is best described by each of the following four scenarios?

SCENARIO 1

Norma, majoring in international and technical communication, has created a portfolio in a folder on her desktop that contains all of her creative writings and drawings. Her portfolio also contains a PowerPoint presentation about Mexico that she created for a social sciences course and a letter written for her business communication class that she has since translated into Spanish. Her portfolio also contains several drafts of an article she is writing for an environmental science class.

SCENARIO 2

Chris is majoring in digital media and will graduate in 2 months. He wants his portfolio to emphasize the high-end tools and skills that he learned in his courses. In addition to his paper portfolio, he will create a CD to showcase his high-end digital media projects. Chris has decided to include on his CD, among other projects, screen captures of a Web site that he created during an internship and streaming video he shot for a multimedia presentation. His portfolio also includes a finished article he wrote for his journalism class and a revised chapter from a software documentation manual that he coauthored.

SCENARIO 3

Sarah is taking a graphics course and must submit a portfolio as her final exam. Her portfolio includes a photo essay, many different business graphics, a creative resume that combines text and graphics, and several pieces that have different typographical elements such as dropped capital letters and pulled quotations.

SCENARIO 4

As a graduation requirement for his architecture program, Mark must submit a design portfolio that demonstrates what he has learned in his 4-year major courses. The portfolio must include a cover, a table of contents, a design statement, and several design projects. It will be reviewed by two of his architecture professors, two architects from a local architecture firm, and a civil engineer. Mark decides to include an Auto Cad rendering of the new Waterloo Visitors Center where he works part-time, a two-dimensional drawing in black and white to show high contrast, several high-quality photos and a three-dimensional model of the Visitors Center, and a four-color marketing brochure promoting the local architectural firms he interned with this past summer.

REFERENCES

Belanoff, Pat, and Marcia Dickson, eds. *Portfolios: Process and Product*. Portsmouth, OR: Boynton/Cook-Heinemann, 1991.

Bishop, Wendy. "Designing a Portfolio Evaluation System." In *Teaching Lives: Essays and Stories*. Logan: Utah State University Press, 1991.

Elbow, Peter, and Pat Belanoff. "Reflections on an Explosion: Portfolios in the 90's and Beyond." In *Situating Portfolios: Four Perspectives*, edited by Kathleen Blake Yancey and Irvin Weiser. Logan: Utah State University Press, 1997.

Gomez, Emily. "Assessment Portfolios: Including English Language Learners in Large-Scale Assessments." *ERIC Digest* (December 2000). http://www.cal.org/resources/digest/0010assessment.html (accessed June 2, 2004).

Kendus, Steven M. "Developing a Web-Based Portfolio." *Intercom* 49, no. 9 (November 2002): 4–7.

Kimball, Miles. *The Web Portfolio Guide*. New York: Longman, 2003.

Linton, Harold. *Portfolio Design*, 3rd ed. New York: W.W. Norton, 2003.

Scott, Cathie, and Carolyn Plumb. "Using Portfolios to Evaluate Service Courses as Part of an Engineering Writing Program." *Technical Communication Quarterly* 8 (1999): 337–50.

University of Washington College of Engineering. "ABET 2000 Information: Portfolio Evaluation Project (PEP)," 1998. http://www.engr.washington.edu/abet/PEP%20Presentation/tsld002.htm (accessed May 23, 2005).

Yancy, Kathleen Blake, and Irwin Weiser, eds. *Situating Portfolios: Four Perspectives*. Logan: Utah State University Press, 1997.

APPENDIX A: SELECTED WEBSITES WITH SAMPLE STUDENT PORTFOLIOS

College of Education University of Florida Student Samples
http://www.coe.ufl.edu/school/portfolio/examples/examples.htm

East Carolina University College of Education Department of Library Science and Instructional Technology Instructional Technology Student Portfolios
http://lsit.coe.ecu.edu/it/Portfolios

Elon University, Elon, NC
Student Portfolios
http://www.elon.edu/students/portfolio/

Humanities and Technical Communication Department
Southern Polytechnic State University
www.spsu.edu/htc/home/undergrad/ug_prog.htm

eFolio Minnesota
http://www.efoliomn.com/index.asp?

Kalamazoo College Outstanding Portfolios
http://www.kzoo.edu/pfolio/outstanding.html

Maricopa Community Colleges: Electronic Portfolio Examples
http://www.mcli.dist.maricopa.edu/dd/eport05/demos.php

Skidmore College Career Services Online Portfolio Center
http://www.skidmore.edu/administration/career/resume/portfolio.html

The Portfolio Clearinghouse
http://www.aahe.org/teaching/portfolio_db.htm

Texas Tech College of Education Digital Portfolio Pilot Project
www.educ.ttu.edu/portfolio

University of South Dakota IdEA Program: Electronic Portfolio Project
http://www.usd.edu/idea/eportfolios.cfm

University of Washington Technical Communication Department
"Technical Communication Professional Portfolio Program Example Portfolios"
https://courses.washington.edu/tc493/examples.html

Wake Forest University: Technology in Education
http://www.wfu.edu/~cunninac/students2k.html

2 Creating a Portfolio Identity

I had a panic attack because I realized I didn't know what I wanted to do or the kind of professional I wanted to be in technical communication—I panicked and my mind went blank. I now had to ask myself the very deep question: What do I really want to do? Norma

I learned that I am a loner who is willing to take risks. That was a big revelation, and it is likely the revelation that will shape my career and life after college. Miranda

INTRODUCTION

This chapter guides you through the process of creating an identity and professional personae for your portfolios. You will explore this notion through heuristics and personality tests that ask you to define your particular niche within the field, along with developing your own sense of style and direction. You will also work to define the rhetorical context for your portfolios as you assess your purposes, audiences, and strategies. Chapter 2 covers the following topics:

* Establishing a professional identity
* Researching and assessing personality
* Shaping a professional identity
* Understanding the rhetorical situation of your portfolio

ESTABLISHING A PROFESSIONAL IDENTITY

What does it mean to establish a professional identity? For many years you might have come to know yourself solely in the role of student. Your identity is wrapped around classroom assignments and professor and degree expectations. It is tied to your relationships with your classmates and the achievement of your degree. We recognize that as a student in a particular program, it is natural to tie your work to particular classes or people. This is not only true for students. As a working professional, you also might see yourself in terms of your current job rather than in terms of your career goals. You might have worked on your current job for many years but are now looking for a way to redefine yourself in the job market, or you might have acquired new skills and want to seek new challenges. Creating a portfolio helps you to articulate your goals and directions. It gives you the opportunity to control the professional identity you want to create for yourself in light of your needs, skills, and personal expectations. It also allows you to present yourself as a distinctive individual within a competitive job market. Almost everyone in communication fields has samples of writing and graphics skills, but your portfolio should also reflect who you are as an individual and demonstrate a sharp sense of audience awareness.

RESEARCHING AND ASSESSING PERSONALITY

As the chapter-opening quotations show, your personality is an important element in determining your career direction. Before you can shape your professional identity, it makes sense to reflect on this dimension of the process. For example, start with the following questions:

- What do you like?
- What do you dislike?
- How do you see the world and how do you interact with other people?
- What do you prefer in terms of variety, security, risk, and job location?
- Do you prefer to work in teams or on your own?
- Do you like to travel or do you have needs that demand flexibility in terms of time on the job?

Some people are very happy as entrepreneurs because they see themselves as extroverts who are very comfortable with risk and cold calling on potential customers. Others enjoy the challenge and security of working through a national company because of its track record, benefits, and a steady paycheck. Still others like the idea of daily interaction with people, while others prefer working primarily on their own. Through our work, we have met all kinds of students and professionals who bring their individual personalities to their work choices.

Identity is constructed through how you see yourself and how you are seen by others. Much of what you know about your personality is gathered through your accumulated responses to life experiences. You can also gain valuable insight into your personality through the perceptions of others. A good way to start your thinking in this area is to list some of the traits you would use to describe yourself. Then, to complement this activity, you might ask a few people who are close to you (family members, friends, coworkers) to briefly describe your personality. Often these two perspectives will match, but in some cases you will see a disparity between your own description and the descriptions of those around you. It is not uncommon to find that, for example, you consider yourself unorganized but others see you as quite the opposite. It is important to compare these perceptions to get a clear idea of how you really feel and to understand which areas you want to emphasize or deemphasize as you make career choices. Many people place themselves in situations that are the direct opposite of what they really want, setting themselves up for misery, failure, or both.

We asked our students to take a close look at their personalities to begin connecting them to their career choices. There are many online sites that can help you understand your personality. They generally take you through a series of questions and help you place your personality in different categories. As a starting point, you might try some of the more popular psychological personality type indicators such as the Myers-Briggs Type Indicator® (www.myersbriggs.org) or the Keirsey-Temperament Sorter® (www.keirsey.com). These tests often break personality down into categories and show tendencies such as extroversion or introversion.

The Keirsey, like the Myers-Briggs, categorizes four temperaments and classifies people as "Rationals" "Idealists," "Artisans," or "Guardians." Each of these categories has characteristics related to planning, problem solving, and worldview. Other instruments assess values, attitudes, or your patterned responses to particular situations. Some tests draw upon your responses to images, and others focus on specific issues such as career choice, self-esteem, power, or communication.

A simple Internet search using the term "personality tests" will present a range of choices to consider. Look closely at the sites you discover; they will often refer you to other useful sites for analyzing your personal and professional identity. Although these assessment tools are not always reliable (some of them are not even serious), they still provide a starting point for productive reflection, allowing you to begin to locate yourself with the responses. Even if you feel that they miss the mark in terms of how you see yourself, you can shape a sense of identity as you respond in relation to or against their profiles.

SHAPING A PROFESSIONAL IDENTITY

We asked our students to use this kind of reflection and go to several online sites that helped them begin to understand their worldview and temperament. What follows is a series of examples that might help you conduct a similar personality analysis as a way of understanding your career and portfolio directions. We start with Miranda's observations:

> My results from the Keirsey Temperament test told me that I believe in basic principles—the dichotomy of good and evil and that I am willing to act on those beliefs. It also told me that I am extroverted and anxious to share what I learn with others. Further, the Career Interest Profiler told me that I'm willing to take risks, which played out over the semester and throughout the creation of my portfolio. The greatest indication of my willingness to take risks (which I never recognized before this class) is the career goals I discussed early on with my mentor—freelance work, mainly followed by graduate school and then teaching. All are risky choices fresh out of college.

As Miranda reflects on her relationship with risk—a personality trait—Tom focuses on how he likes to spend his time and how he relates to other people:

> I like to write sometimes; I really have to be in the mood to enjoy it. The moods are coming more often. I definitely like learning new things. I am good at avoiding conflict, usually. I think I can negotiate compromise. I'm excellent at graphics. I'll often make an executive decision to answer something no one else will. I probably don't take criticism well, but who does? I do get upset when the framework changes, but I do comply with the changes. I always keep my time commitments.

Notice the way that these students begin to connect their personalities with particular career choices. Another student, Brian W., sees himself as a "warm person" who wants his work to have a personal dimension. Here he makes connections between his personality and his possible career choices:

> I am a warm person who even in a professional setting tries to bring everything back to a personal level. I like to be where the action is and do not like to sit on the sideline long. I am a doer, not one to stand by and watch. The test results also mentioned that my personality finds [that] doing something that isn't fun or enjoyable isn't worth doing. Today must be enjoyed because tomorrow may never come. I see myself as a professional who does work more on an artistic level than one who is punching a clock, being handed work, told how it should be completed and then punching the clock again at the end of the day. I like the multimedia work, but I do it with my mindset—my way.

His career choice of multimedia and video production will allow him to jump into the action and get off the sidelines. His statement that "Today must be enjoyed because tomorrow may never come" indicates a tolerance for ambiguity and a love of immediacy. In comparison, look at the way Amy's focus on writing connects to her career choices. She states:

> I am a hard worker who likes order. I like things presented in a straightforward manner with pure factual information to back it up. I keep a cool head in high-stress situations. There is very little that can upset me in the workplace. I am dedicated to my work. I like challenges. I am picky about how my work turns out—I like neatness and logical presentation of material. I am a stickler for details. I like simple, factual evidence. I am honest and straightforward, so I have no qualms about saying what I think about how a project is proceeding.

Amy likes to deal with facts and enjoys order. Notice the differences between this more highly structured person and Brian W., with his desire for autonomy and variety on the job. Amy pushes this connection even further as she gets specific about the kind of job she sees herself in after she graduates:

> I see myself in an entry-level job when I first graduate, working my way up into a higher position as time passes. I want to work for a medium-sized company where I can get to know people yet still have the opportunity for upward movement.

Amy seems to desire job security with a stable company, clearly accepting the fact that she will have to work her way up and start with an entry-level job. She shapes her expectations in anticipation of this kind of career movement. She has decided that she wants a medium-sized company—one that is not so small that it is unstable but not so big that she gets lost in the corporate setting.

Miranda does a good job of connecting the knowledge gained from the personality tests to her professional identity. Using these tests, she starts to refine the career objectives she wants to demonstrate through her portfolios:

> According to the Motivational Appraisal of Personal Potential, I am motivated to describe, explain, teach, illustrate, and interpret. This is a journalistic trait dedicated to informing people. I am a teacher at heart, but I don't need a lot of money or fame to be happy in my career. And, strangely, though I like sharing information, this test confirms that I do not necessarily need interaction in the workplace. The Keirsey Temperament Sorter tells me that I am inclined to seek out the truth and disclose that to others, making me tireless in conversing with others. Keirsey also tells me that for me, "nothing occurs which does not have some deep ethical significance." Surprisingly I learned from my Career Interest Profiler that I am willing to take risks in my professional life in order to ensure a professional life that offers variety.

These comments demonstrate Miranda's desire to teach and inform but also hit upon an important ethical dimension of her personality that might lead her to different career choices. She likes to talk, interact, and help others, and she has a strong desire for variety in the workplace.

Wylie used the personality tests to come up with the following ideas about his professional personality:

* I am creative.
* I can make things better.
* I don't approach problems in the same way that others do, and that's a good thing.
* I like to color outside the lines and think outside the box.

Although he comes to understand these things about himself, Wylie explores his personality a bit more deeply to consider possible misunderstandings. This is important because there is more at stake than just assuming a professional personality. You need to reflect and consider contexts as you refine your desired image. The next quotations from Wylie's proposal show him working to complicate (in parentheses) these ideas to find an appropriate professional identity:

Possible Misunderstandings

* Jack of all trades (Gamble hiring me)
* The 3M slogan: I don't make things, I make things better (Won't make anything)
* Racing theme (Race through work and not do it right)
* Outside the box (People like it inside the box, it's safe in there)

We see him productively working through his ideas to get a better sense of how his audience might perceive him. Wylie summarizes some of his priorities. He says:

> I will focus on getting a job in the technical communication industry that involves working with computers. People who work in video production and visual communication use computers frequently in their careers. I enjoy creating videos and unique websites. Boredom is my main fear factor and drives me to search for an exciting career. In video production I would always be working with something new.

Joy, as she refines her professional identity, begins to make important connections between the knowledge gained through the personality tests and her own career direction. After taking the personality tests, Joy has found that she was consistently classified as:

* Very detail-oriented
* A planner
* Organized

* A fact-based decision maker
* A leader

Joy has also found out something about her day-to-day work practices. She has come to understand that she had a high need for goal-oriented activities. She explains:

> I realized that all of my activity must be goal-oriented. I didn't realize this before, but sometimes I find it difficult to participate in leisure activity when I don't see an accomplishment at the end.

Although Joy recognizes her "admirable traits for a professional writer, editor, information designer, or leader in any corporation," she also reflects on how some of these same traits might be viewed as weaknesses:

> On the downside, I am so goal-oriented that I will make decisions based on fact rather than the feelings of others. I will readily discard personal interaction when in pursuit of a goal. I also tend to be abrupt when I don't feel that the verbal exchange contributes positively to my agenda. This may be a problem for me in the office politics realm.

Joy moves from more open-ended reflection to summarizing some of her results and includes a quotation from the profiles she deemed significant:

* *Realist, fact-based, analytical*—"prefer that decisions be based on impersonal data, wants to work from well-thought-out plans."
* *Witty*—finds humor in almost every situation.
* *Creative*—talent for sketching and likes color, music, and literature; gets bored with repetitive tasks.
* *Efficient, organized, productive*—"good at ordering priorities, generalizing, summarizing, and demonstrating ideas."
* *Scheduler*—"plans in advance, keeps both short- and long-term objectives well in mind."

Next, she connects them to particular career interests:

* Educational/instructional: corporate trainer, vocational school instructor
* Research/analysis: archivist, journalist
* Writer: corporate communicator, journalist

She takes this one step further and begins to define areas of specialization:

* Experience in the printing industry
* Analysis
* Organization
* Graphic design
* Public speaking

Finally, she incorporates some of the goals and practical expectations she has for a potential job and lists the following:

* Job with benefits
* Outlet for creativity
* Flexibility
* Earnings sufficient for lifestyle
* Work until 55 years of age

Through this work, Joy comes up with the following statements and priorities:

> * I have excellent organization skills and pay attention to detail. I also love to read. These strengths are critical for editors or information designers. I have recently revealed a talent for drawing and want to somehow incorporate that into my career.

 ❋   I don't wish to travel, relocate, or work longer than 8 hours per day.
 ❋   I can see myself as an editor for a magazine, periodical, or publishing firm based in the South. My role
     would be mostly autonomous, requiring periodic group meetings and projects that would require reporting
     to an office setting on a periodic basis.

Another student and working professional, Wilda, used this reflection to reinforce her attraction to her current job (in the Centers for Disease Control and prevention [CDC] communications department) and to help her consider potential directions to pursue. She says:

> While in my professional life I am loyal and methodical, I don't work well in large groups; I prefer small
> groups or one-on-one situations. I don't like being in a supervisory role. I prefer to work alone and I am bad
> about delegating work because I believe that "if you want it done right, do it yourself." I am organized to a
> fault. I form personal bonds with people, not institutional bonds. I have a hard time accepting praise for my
> accomplishments, but I take pride in everything I do. I don't like to call attention to myself because it is part of
> my job to do the job correctly in the first place.

Wilda has done a good job of assessing her professional personality including her work ethic, interpersonal style, and work environment. Although she is very happy with her current job, she desires a position in which she could "do more research and develop procedure manuals for different areas." Through this exploration she has discovered that her overall goal is to work in the area of knowledge management as a librarian. She explains:

> Working as a librarian would fulfill most of my goals. I could work with books, which I love. I can research
> and document information and work quietly and undisturbed most of the time. Cataloging is good because it
> puts items in order. I always like to have everything in place.

Notice how this last statement shows Wilda pulling together her personality and skills to create a professional identity while at the same time moving to a tangible career choice that draws on both.

All of these elements must work together to create professional identity: personality, goals, context, skills and talents, and expectations.

EXERCISE 2.1 PERSONALITY RESEARCH/ASSESSMENT

This assignment asks you to take a close look at your personality through research and analysis of the results of online personality and career tests. You might start with the Keirsey Temperament Sorter http://keirsey.com/ and then go on to find other sites that address both your personal and professional personalities. The search terms "personality tests," "personality tests and careers," and "career tests" will all lead you toward these sorts of assessment instruments. Try to hit professional as well as personal sites. Briefly describe each site and what you learned from it (include full URL documentation). Work to connect the personal and the professional.

EXERCISE 2.2 PROFESSIONAL IDENTITY: NARRATIVE

In a single page of exploratory writing, describe your professional identity. Use the information you gathered through the assessment and heuristic. Describe your most compelling traits and skills in relation to your career direction. Work to be as specific as possible as you examine the different dimensions of professional identity (skills, life situation, location, company size, personality, motivation, etc.). You will use this statement to begin shaping your portfolio.

EXERCISE 2.3 PROFESSIONAL PERSONALITY HEURISTIC

Address each of the following questions in writing (provide a couple of sentences for each response).

- Briefly describe your personality (based on assessment tools and self-knowledge).
- How do you see yourself as a professional?
- How do you see yourself fitting into the technical communication/professional writing profession?
- What kind of work do you most enjoy?
- What kind of work do you most dislike?
- How do your particular skills and desires fit in with the larger picture of your life?
- Are there any particular life situations that might affect your professional direction?
- In what kind of company, role, and position do you see yourself?
- How does your personality fit your career choice?
- How do you want to spend your time at work?
- What is your ideal job (be realistic)?
- List and prioritize the qualities you desire most in your future work life (travel, autonomy, variety, flexibility, creativity, money, room to grow, etc.).

UNDERSTANDING THE RHETORICAL SITUATION OF YOUR PORTFOLIO

As a communication student, you have come to understand that every written document, graphic, or multimedia piece comes from a particular "rhetorical situation." Although this term was used by ancient rhetorical theorists such as Plato and Aristotle, it was more recently coined by the contemporary writing theorist Lloyd Bitzer (1968) to refer to the ways we analyze context to create effective rhetorical discourse. The rhetorical situation is based upon the choices that we make as communicators in relation to our own **purposes, audiences,** and **subjects.** We do this naturally in our everyday lives as we change our language to speak to our families, friends, and employees or coworkers or for different purposes such as persuasion, humor, or information. For example, your rhetorical situation will shift both your purposes and your audience as you move from an academic to a professional setting. As communicators, our goal is to carefully assess each rhetorical situation and come up with the best language and approach for it. The same is true for your portfolios. Consider them as a form of communication that works within a particular rhetorical context. Before you design your portfolios, it is important to reflect on and articulate the following rhetorical elements.

Rhetorical element 1: purpose

This element deals with your intentions and purposes. You need to figure out what you want your portfolio to express. Your work on developing your professional identity should begin to shape your ideas about purpose. Purposes might include defining your personality traits, skills sets, or job structure. This might include communicating a particular niche or identifying a direction in which you would like to head. We asked our students to extend their work with the personality profiles to understand their purposes, drawing on their immediate and long-term career objectives to help determine the purposes of their portfolios. For example, Tom states:

> I want to put an emphasis on my graphic art skills. I will include my creative resume, a set of instructions, a brochure, a hardware manual sample, a magazine article, and other materials demonstrating an integration of illustration and graphic arts skill.

This comment shows him emphasizing his skills and defining his particular talent for illustration. Tom, who collected materials (such as brochures, illustrations, and flyers) from former consulting jobs, also decided to include a section showcasing his talents. He has transformed this sense of his individual niche—graphic art skills—into artifacts he might include in his portfolio. These choices directly connect to his purposes, letting potential users/readers recognize his particular specialization.

Sometimes purposes involve personality and life circumstances. Miranda hits on both of these issues as she tries to balance her desire for freedom with the reality of her financial situation. She reflects:

> There is really only one way to describe my life situation—free. I have nothing holding me back from moving out of state, out of country. I am, however, operating without a safety net. I am not financially secure and have no one to turn to for financial help. So I will have to carefully consider any risk I decide to take.

There are many variables that establish purpose. Although we suggest that you shoot for the sky, we know it is unrealistic to think that all of us—or any of us, for that matter—can start at the top. When shaping your purposes, it is important to consider both the short term and the long term, along with your ideal and practical ideas about your career's direction and focus.

Rhetorical element 2: audience

In many ways, it seems obvious that our language and methods of communication change as our audience changes. For example, we speak very differently to a four-year-old child than we do to his adult counterpart. By the same token, you will encounter many language and design changes when you move from an academic to a professional audience. You will need to determine the main audience for your portfolios. This might mean determining a particular niche or specialization or presenting yourself as a generalist with broad-based flexibility. If you are entering the design field, you might consider an audience that is open to creative presentation and subject matter. Some designers included "creative resumes" and original artwork to appeal to this particular audience. On the other hand, if you are hoping to enter a government agency or the medical field, you might prefer a more conservative approach. For example, Michel, who wanted to work for the CDC, a government agency, included documents with medical content and language that demonstrated her knowledge in this specialized area. She also used a conservative color scheme and design that would be considered "classical" or "serious" rather than "artsy." Another student, who wanted to use his communication skills in law enforcement, chose to stick with the documents in this field and also had to consider a design theme that respected its hierarchy. He chose the classic blue and gold along with other design features, but he was continually concerned that they might be misinterpreted by his audience (as lofty goals of achievement outside of rank). There are many issues that determine the audience. In fact, some students created more than one portfolio to present in different professional settings.

Before moving too far ahead in this process, you should carefully analyze your audience. The following comments reflect our students' attempts to analyze their audiences to help them choose a direction for their portfolios. Some students saw their audience in terms of a particular job or professional niche. For example, Sarah, who wanted to shape her portfolio for a career in marketing, envisioned her audience as advertising, marketing, and public relations agencies. Brian W., as he suggested earlier, targeted an even narrower audience, focusing on multimedia productions with an audience in the film and video production industry. Other students, such as Amy, preferred to present themselves as generalists in which multiple audiences might find a fit within a more diverse range of materials. Rather than shaping her portfolio for a particular specialization, Amy included a little bit of everything so that she would be seen as flexible and versatile. Some students created multiple portfolios, depending on the kinds of jobs they planned to pursue. Your audience will change in relation to your area(s) of concentration.

In Miranda's statement, we see her searching to find her audience as she talks about the general "users" of her portfolio:

> I also put a lot of emphasis on explaining to the users what they were seeing. Contextualizing each piece took a lot of consideration, and I worked on it throughout the last half of the semester. I considered organization, and in cases where my work was unbalanced, for example my journalism section, I struggled to find a way to show both the depth and breadth of my experience without overwhelming the users.

If you plan to concentrate on a particular area or professional niche, we still recommend that you include at least some materials that demonstrate a range of skills. After all, you might be one of several applicants who all have graphic skills, but your additional writing or teamwork skills might give you an advantage over a similar candidate.

The audience also shapes your intentions and personality. For example, Miranda, in her desire to showcase herself as a risk taker, might take some risks in the design of her portfolio. Although she wants to appear professional to her audience, she also wants to show that she is an innovator with a unique vision. Michel, who wanted to work for a government agency, knew that a more conservative approach would be more appealing to her audience due to the sensitive nature of the material she would be writing. Brian W., wanted to incorporate his hobby of car racing into his multimedia job and made choices that connected him to this particular community. Norma wanted to emphasize both the skills of her major and her Spanish and Social International Studies minor while at the same time working in a less technical capacity:

> I want to market my skills for the professional business end of technical communication and not so much the technical end. I want to have a career where I am allowed a certain degree of creative freedom. I want to market myself as a versatile technical and professional communicator. I want people to see that I can adapt to different situations.

Rhetorical element 3: subject as representative skills

Communicators traditionally think of the subject as relating to the topic, viewing the content of the communication in relation to the other elements. In many cases, your choice of content will be based on your specialization such as journalism, medical writing, or editing. However, when determining the content or subject of your portfolios, it might also be helpful to think of this element in terms of skills. In essence, the content of your portfolio is determined by the skills you hope to represent.

There are so many opportunities in the fields of technical and professional communication that it always makes sense to demonstrate, through your portfolios, the kinds of skills you offer. You can use your portfolios to draw together these skills. Some of them will be specific skills such as particular software applications, design principles, or writing formats. Others will be less tangible skills such as organization and teamwork. A quick scan of job ad descriptions will help you to determine the skills that are important in your field. The next sections of this chapter present examples of some of these skills sets that you might showcase through the content or subject matter of your portfolios.

CATEGORIZING REPRESENTATIVE SKILLS

Although there are many skills you might want to represent, it is helpful to start categorizing them into the following major divisions. Although this list does not cover everything, it gives you another way to start dividing your work and defining your professional identity.

Technology skills

In today's competitive workplace, employers are looking for individuals with technical skills. Writing and communication, once viewed as central professional elements, are enhanced by integration with technology. This means everything from word processing skills to document design to knowledge of specific web publishing applications such as Flash or Dreamweaver. Technology is also an excellent way to present particular skills from an academic minor area or concentration such as computer science or business. Think about what skills you bring to the table. Do not hesitate to include technical knowledge even if you feel it is obvious, as you never know what people need. It works in your favor to create the impression that you are familiar and comfortable with adapting to and learning new technologies. On the other hand, be realistic. Do not state that you have mastered a complicated application such as Frame Maker when your program skills are minimal. Do not assume that you will have the time to learn. If you feel that there are gaps in your technical knowledge, learn the software before you mention it in your portfolio. It is a great idea to use the creation of new portfolio pieces as a way to increase your knowledge of and comfort with a particular software application.

Graphic and design skills

Think about the graphic skills you have used in your work. You might start by reviewing your work and listing these skills. Do you know how to do desktop publishing? How have you used colors and shapes and demonstrated design sensibility? Do you have photography skills you wish to show or original drawings and illustrations? This area also includes document design skills such as layout, font knowledge, and incorporation of visual images.

Communication skills

What do you want to say about your communication skills? First, break them down into the language skills involved: reading, writing, speaking, thinking, and listening. Whether or not you include them in a job description, these are the less tangible things that impress others and help present you as a communicator. As a student of communication, you are expected to have all of these skills.

We will go through them briefly, one at a time.

Written communication.　This is the most obvious skill, since you have probably written many assignments throughout your degree program or generated a variety of writing projects on the job. It includes everything from memos to annual reports, advertising copy, Web site content, and press releases. What skills are most important for writing? How might you show your ability to generate ideas, draft, revise, and edit?

Analysis and research.　How do you go about analyzing a problem? Are there ways that your work demonstrates your ability to think and learn? Look for places where you might discuss your thinking processes. You might describe your research skills and show your reports, manuals, and proposals.

Oral presentation and speaking skills.　In the workplace, you will often need to present your work. This begins with your individual presentation and the ability to speak articulately about your subjects. You will also be required to make formal presentations or speeches. Speaking obviously occurs in many different contexts, like conducting meetings and performance evaluations. Your spoken presence will be the first thing employers and clients notice as you interview for jobs or compete to get their business. You might also demonstrate your knowledge of presentation technology by using PowerPoint presentations and graphic charts.

Organizational skills.　In the modern world of multitasking and through the extension of technology, generally the most organized people are the most successful. Employers appreciate a strong sense of organization because it is connected with trust. Your portfolios themselves will demonstrate organization, providing a collection of work pulled together neatly for viewing. Think about the way your whole portfolio is organized so that it is easy to read and has a definite progression of order. You can also show organization by including projects that you managed, preproduction sheets, and training manuals.

Teamwork and interpersonal group skills.　Another desirable skill is the ability to work well with others. This might involve leadership, collaborative writing, understanding of interpersonal relationships, and the ability to manage collaborative projects. Companies or clients want to know that you will fit in and be an asset to their team. They also want to know that you have the skills to handle conflict, make sound decisions, and motivate others.

Table 2.1 summarizes the connections between skill sets, specific skills or traits, and examples of artifacts.

TABLE 2.1
Connections between Skills and Artifacts

Skill Sets	Specific Skills/Traits	Artifacts Examples
Technological	Software applications, Web design, video production	Web sites, online design, PowerPoint presentation, multimedia film
Graphic	Design sensibility, color, document design	Brochures, Web sites, logos, marketing campaigns
Written Communication	Clear style, editing, rhetorical awareness	Essays, articles, reviews, brochures
Research	Analytical and interpretive	Research projects, data analysis
Presentation	Organization, spoken presence, awareness of technology	PowerPoint presentations, scripts
Organization	Reliability, order, flexibility	Compiled portfolios, training manuals, project management reports
Teamwork/Interpersonal	Leadership, compatibility, project management	Team projects, training manuals, certificates

Rhetorical element 4: context

Context refers to all of the outside variables that might affect the development of your portfolio. Think about what you need to know to create a portfolio that fits the current culture of your field and the job market as a whole. You will need to analyze what is considered state of the art in terms of philosophy, design, and technological applications. We also ask you to consider issues such as the economy and other trends and conventions that might influence the reading of your portfolio.

In order to determine context, you might read job announcements, talk to other professionals, or read journals or trade publications in your field. You can also look at market trends and analyze other external influences through the media.

SUMMARY

As you think about the professional personality or the rhetorical situation, you should move toward shaping a professional identity for your portfolios. This reflective work should guide you through the rest of the process as you make decisions regarding design, content, and theme. It is important to understand yourself and your goals before communicating them to others. This conscious reflection will engage you in productive invention as you generate ideas and create a direction, focus, and identity for your portfolios.

ASSIGNMENTS

Assignment 1: Skills Inventory

Conduct a skills inventory in which you list all the skills that you believe you have at this time. To go beyond your initial list, review your artifacts to remind yourself of what was involved in their creation. Broaden your scope as you include all the skills described in this chapter (writing, interpersonal skills, knowledge of technology, etc.). You might also categorize them by skill type, rank order, or preference.

Assignment 2: *Analyzing Your Rhetorical Situation*

Using the elements described above (purpose, audience, subject/skills, context), analyze in writing the rhetorical situation of your portfolio. Briefly explore each aspect to create a rhetorical profile for your portfolio.

REFERENCES

Bitzer, Lloyd. "The Rhetorical Situation." *Philosophy and Rhetoric* 1 (1968): 1–14.

Keirsey-Temperament Sorter. www.keirsey.com (accessed August 15, 2005).

Myers-Briggs Personality Type Indicator. Gainesville, FL: The Myers-Briggs Foundation www.myersbriggs.org (accessed August 15, 2005).

3 Portfolio Contents, Design, and Structure

Although I was stressed out all semester over my portfolio, I felt okay with the progress because everything was planned. I knew where I was going and how I was getting there. Miranda

INTRODUCTION

As you can see from this quotation, the process of organizing your portfolio can be daunting. As Miranda suggests, it was her preplanning that helped ground her as she moved through the process. This chapter helps you come up with a plan that will guide you in making decisions about the content, design, and structure of your portfolios. To help you move toward a carefully considered plan, Chapter 3 covers the following topics:

- Considering contents for the portfolio
- Matching skills with artifacts
- Understanding theme and metaphor
- Creating a portfolio design proposal and table of contents

CONSIDERING CONTENTS FOR THE PORTFOLIO

Now that you are starting to define your professional identity and goals, it is time to think about the contents, or artifacts, to include in your portfolio. We use the term "artifacts" rather than "documents" because technical and professional communication involves representing yourself in forms beyond the written page such as video, audio, and multimedia projects. Initially, the job of collecting and selecting materials for your portfolio may seem overwhelming. You should start collecting these materials as soon as you decide to pursue a major in communication or throughout your employment in related communication fields.

As you noticed in your online portfolio analysis in Chapter 1, portfolios contain a variety of artifacts. In a field as wide and diverse as technical and professional communication, there are many types of professionals. Considering your minors, consulting projects, and particular emphases, you should have much from which to choose (see Chapter 1). Hopefully, as a student or professional, you can now return to backup copies of all your work—a folder on your desktop, a CD, paper copies of your work with teachers' comments, and even photocopies of paper documents.

At this point in the process, you should not be too selective. If you are selective too early, you may end up discarding a document that later on, when revised, could become a major piece in your portfolio. At the beginning of the process, do not rule anything out and try to think creatively about what you already have. This process will also reveal potential gaps and places where you might have to generate new material. Some of you will be reflecting upon the projects and papers you have completed within your degree program; others will go back to artifacts created for particular jobs, including those produced for consulting clients. In many cases, your final portfolio will draw from a combination of these sources. Your goal is to collect as much of your own material as possible related to your professional work, coursework, professional activities (including internships), community service projects, campus activities, or even your personal life (when deemed appropriate).

Portfolios are representative rather than comprehensive

After collecting all of the possible artifacts to include in your portfolio, you will need to return to the work you completed in Chapter 1. One basic principle is that *portfolios are representative rather than comprehensive*. This means that your portfolio will not be just a container for everything you have done. Instead, it should be carefully crafted to address what you consider important. It will be your job to narrow down and choose particular items that best represent your work, diverse skills, and professional identity. Your choices should reveal your values and goals and be directed to a particular audience. Your main purpose here is to demonstrate what you can do and who you are as an individual and as a professional.

Remember, you might have a limited amount of time to display your portfolio, so strive for balance and range within your representative pieces. Avoid having many documents that do nothing more than duplicate each other. Instead, look at the similarities among your artifacts and choose the best pieces demonstrating each skill or trait you hope to represent.

Artifacts and contents of your portfolio

What types of artifacts are found in portfolios? Remember, at this point in the process you are still pulling everything together, casting a wide net. This means writing, graphics, and media projects along with projects that represent other, less tangible skills such as collaboration, organization, leadership, and innovation. You should collect all the items produced within your degree program, from former jobs, or for consulting clients.

There are a number of ways to organize and present the artifacts in your portfolio. As a general rule (as stated in Chapter 1), your paper and electronic portfolios will probably contain most of the following elements:

* Resume (with a pocket in a paper portfolio holding extra copies)
* Welcome page or homepage
* Table of contents
* Context statements: a short analysis of each document or section noting what it is, why it was included, its audience, the tools used to create it, and the skills it represents
* Professional writing/communication samples grouped into categories or skill areas
* Awards and other forms of recognition
* Professional certificates
* Letters of recommendation

Beyond this general description of the contents, it is up to you—to choose artifacts that project your skills and professional identity. Table 3.1 shows some artifact genres along with possible samples of each type. This list of suggestions, although not all-inclusive, should help you see the range of possibilities.

MATCHING SKILLS WITH ARTIFACTS

To collect a representative sample, it is important to define clearly the skills and objectives you want to communicate. After reviewing your artifacts, you might determine that you need more balance. Possibly you have an overabundance in one area and a deficiency in another. Many projects might appear unfinished or insignificant. It is time to start choosing, discarding, and shaping your contents to match your skills. Assume that you will revise and edit most of the documents you choose to achieve portfolio-quality work.

TABLE 3.1
Artifact Genre and Samples

Artifact Genre	Artifact Samples
Business and Professional Writing	Memos, letters, resume, annual reports, team projects/ presentations, marketing materials, proposals, newsletters, style manuals, emails
Technical Writing and Documentation	Edited texts, tables/charts/graphs, training materials, users guides, software documentation, technical reports, business plans, policies and procedures, indexes, information architecture plans, medical writing pieces
Journalism	Editorials/op-ed pieces, press releases, articles, published material, book reviews, speeches, public relations efforts
Academic Writing	Research papers, essays, lab/scientific reports, creative writing (poetry, short stories, plays), drafts and revised texts, oral reports, team projects
Graphics	Logos, brochures, Web sites, design projects, digital design, invitations, CD covers, original artwork, photography, graphic software, illustration and online graphics applications, recommendation and feasibility reports, forms and templates
Multimedia	Video, film, audio, PowerPoint presentations, films, computer animation, musical compositions, storyboards
Web Materials	Web pages, web-based training, online help
Awards/Honors	Certificates, references, honors
Other	Multilingual documents, emphasis areas (minor or specialization), event planning

You should even broaden your focus and think about documents and projects that represent multiple skills and purposes. For example, a brochure can represent skills in writing, document design, organization, research, and audience analysis. It can also demonstrate technological and design skills, incorporating images or design elements (more on this in Chapter 4). Imagine speaking about or explaining your work to others. Which products are you most proud of? Which ones reflect you as an individual? Which ones best highlight the skills you want to emphasize? Reflect upon what you might say about what was involved in each product's creation and what uses it might have in a work setting. Think about how the artifacts cluster together and start to form categories. Begin to visualize an overall structure that includes particular sections for your portfolios.

Addressing employers' needs

Obviously, employers' needs should be a major factor determining what is included in your portfolio (see Chapter 8 for a detailed discussion of the job search), so review several job descriptions in order to gauge employers' requirements. Your portfolio should show that you understand these requirements and have completed projects that clearly match them. If you are applying for a generalist entry-level position, you will probably want to emphasize your versatility by including a wide range of samples in your portfolio. While there is no specific number of pieces to include, an applicant for an entry-level technical writing position may want to include as many as ten to twenty (of course, this number might change as you gain experience in your field). An effective strategy is to place at the front of your portfolio the artifacts that best match the positions you desire or, if you don't have a strong match, open with your best work. If a major job requirement is to work as a member of a project team, you may want to place a group project at the front of your portfolio and include an explanation of its goals, the timeline/schedule, and a clear statement of your contributions. If the job responsibilities emphasize strong editing skills, it may be helpful to place a before-and-after sample at the beginning of your portfolio. Your portfolio should be divided into easy-to-navigate sections providing flexibility to move around, skip sections, or reorder your priorities as you show your work and emphasize particular skill areas. Flexibility will also come into play in your electronic portfolio, giving users the opportunity to choose the order of viewing based on their needs and interests. Your portfolio should stand as a whole body of work, but each section should also be able to stand alone or be read out of order.

Legal and ethical selection of materials

Although these issues are covered in detail in Chapter 6, they are worth mentioning here as you pull together the contents of your portfolios. Make sure that all of your materials follow the legal guidelines for document citation, appropriate use of symbols and images, and permission statements. This is particularly important when you are posting information on the Web. Check your documents carefully to ensure that they meet legal and ethical standards so that you can stand behind your portfolio with professional integrity.

For example, if you choose to include group projects, you must be careful to create an ethical presentation. Make sure that you have written permission from your team members and accurately portray your own role. Group projects are often misinterpreted because they involve collaborative efforts that make individual contributions hard to distinguish. Remember that other team members also have the right to use the same project to represent their skills. In today's competitive workplace, it is not uncommon for employers to interview multiple candidates for the same position, which could lead to embarrassing situations that can be easily avoided. This and other ethical dilemmas will come up as you gather materials for your portfolios.

Exercises 3.1 and 3.2 will help you start collecting and organizing your artifacts.

EXERCISE 3.1 ARTIFACT LIST

Gather and list all of the artifacts that might be included in your portfolio. This is the time for you to broaden your scope and include all possibilities. Start by writing down all of these artifacts in a comprehensive list. Print the list and add the following codes:

- A ✓ next to those items you definitely want to use.
- The letter R next to those items you want to use with revisions.
- A ? next to those items that you are unsure about.
- An X next to those items that you want to discard.

Start a handwritten list at the bottom indicating some items you might need to generate.

EXERCISE 3.2 CATEGORIZING AND ORGANIZING YOUR ARTIFACTS

Once you have finished coding your list, go back through it and start grouping the artifacts. Begin to categorize and organize your work into sections that represent your skills, theme, or genre to start shaping a sense of overall structure. First, create a planning document and then begin placing your contents in a three-ring binder to visualize the order and shape of your portfolio. Place copies of documents in sheet protectors and use tab dividers to begin forming the sections and the order of your portfolio. This version—your working portfolio—allows you to change the contents, order, and categories as you progress toward your final version.

Visual elements

There are many possible ways of adding visual elements to your portfolios (see Exercise 3.3). As a composer and designer, you should consider issues such as color, fonts, graphics, and document design. For the paper portfolio you will need to make decisions regarding, design, layout, paper, and headings (see Chapter 4). The electronic portfolio should incorporate navigation, animation, and online graphics (see Chapter 5).

EXERCISE 3.3 VISUAL ANALYSIS

Return to the portfolio sample analysis you conducted in Chapter 1. This time, as you review the portfolios, focus on visual elements such as font, color, design, navigation, arrangement, and so on. Think about these elements as you review your own work.

UNDERSTANDING THEME AND METAPHOR

Since we emphasize that a portfolio is more than a container for your work, we encourage you to find ways to pull your artifacts together to achieve continuity. Your portfolio should represent who you are and how you want to present yourself to others. This means creating connectivity between your artifacts and their sections. Some portfolios depend on design and color to pull them together. Others are content-driven, held together by subject matter. Still others employ a theme or metaphor as a way of creating consistency and a sense of individuality. Metaphors are sometimes understood only in terms of "poetic imagination." However, as stated in the well-known book *Metaphors We Live By,* linguistic theorists George Lakoff and Mark Johnson suggest that "the metaphor is pervasive in everyday life, not just in language but in thought and action. Our ordinary conceptual system, in terms of which we both think and act is fundamentally metaphorical in nature" (Lakoff and Johnson 1980, 1). They speak of this "conceptual system" as a collective way of thinking that involves connecting ideas to other ideas through images, language, and symbols. This conceptual system, they point out, often operates at a subconscious level and informs all aspects of everyday life.

   This concept is particularly important for communicators who are trying to present not only their generic skills but also their professional identity and distinctive traits. A theme or metaphor can represent particular philosophies, goals, or interests. It can encourage your audience to read and connect to your work consciously and subconsciously. At first, many students resist this idea because they feel it is limiting or too "artsy." Many believe that they are not creative enough to think metaphorically. However, time and time again, we have found that the most solid, interesting, and effective portfolios generally involve a thread that runs consistently throughout. After all, as we stressed earlier, your portfolios involve an intense act of composition that draws on all the elements of your particular rhetorical situation. Choose a metaphor or theme that is unique to you and will leave a memorable impression on your audience.

Our students chose a variety of themes based on their personalities, goals, and skills. They reported that although their theme was often the most difficult element to choose, once they found it, it helped pull together the units of the portfolio into a comprehensive body of work. Table 3.2 summarizes some of the thematic types and samples described in the following subsections.

Theme through professional personality

Many of our students used the work generated in their earlier writings on professional identity and incorporated these personality traits into their theme. For example, Miranda defined her theme through design elements and ideologies from the Art Deco period—a period she equates with her professional personality. She states:

> I'm going to have to work on giving it a continuous look while presenting it to emphasize diversity. Stylewise . . . I want a classic, maybe even a glamorous look that can attract sophisticated clients.

She takes us through her process as she continues:

> I'm thinking about old Hollywood glamour. Glitzy . . . but with clean lines and balance. I want practical . . . not plain. Eye-catching, but not overwhelming. Light silvers . . . blues . . . black . . . monochromatic? A monochromatic color scheme would give the portfolio the continuity I am looking for.

Miranda connects these ideas to a particular historical time period in which art, architecture, and industry were in a state of change. She wanted to communicate the idea that she is a bold risk taker on the cutting edge. She conducted research and immersed herself in the images of the Art Deco period in order to come up with her metaphor. Miranda reported going to Web sites about the period and printing images that she then hung in her room and carried in her backpack so that she would see them all the time and reflect intensely on her subject. She titled her design proposal "Art Deco Portfolio: Industrial-Age Style for the Information Age" and used keywords such as "aerodynamic," "streamlining," "simple format," "clean lines," and "vivid colors," noting that the ideology of the Art Deco period also reflected her ideas and her professional identity. This time period drew from her earlier work on personality, in which she wanted to make a bold statement emphasizing her ability to take risks. As she comments in her proposal, the time period "reflected an unbridled enthusiasm for what man and technology could do" and recognized that the three separate art movements and the 1925 Exposition International promoted a celebration of "living in the modern world." She drew inspiration from the images of the period such as posters, paintings, buildings, and statues. Notice how, in Figure 3.1, Miranda followed through on her decision to use clean lines, stylized fonts, and design elements in her portfolio.

TABLE 3.2
Theme or Metaphor Types and Samples

Type of Theme or Metaphor	Theme or Metaphor Samples
Professional Personality	Traits: bold, risk taker, determination Goals: desire to travel, be an entrepreneur
Personal Interest	Running, race car driving, home improvement
Conceptual	Puzzle pieces, metamorphosis (butterfly)
Symbol	Signpost, electrical circuitry

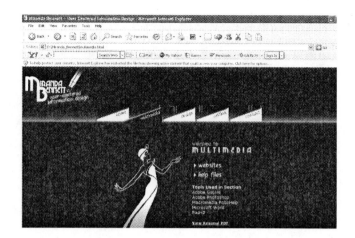

FIGURE 3.1
Miranda's Home Page
Source: Used with permission, Miranda
Bennett 2005.

Sarah also created a distinct impression through her portfolio as she evoked a theme that focused on travel and entertainment—two things that she wanted to incorporate into her career. As she says:

> I would like to use my portfolio to show future/potential employers how my love of travel and entertainment of all kinds could benefit their company.

Her next comment suggests the ways she might tie these themes to the portfolio as she begins to connect to design elements and images:

> I want to use a retro, black background style. Something very unique. I want to use a "movies" theme. The contents could be opening credits. My writing portion could be scripts and the pictures would be the movie part.

Sarah ended up incorporating the travel and entertainment part of the theme by using a different major city for arts and entertainment (New York, Paris, etc.) to head each section. This made sense for her, given her emphasis on international technical communication. In her electronic portfolio, she carried through this theme by using suitcases as part of her online navigation scheme. Her message is definitely communicated, and even if she does not end up traveling or working in the entertainment industry, her potential employers will see the sense of adventure and cultural awareness that are part of her identity.

Figure 3.2 shows the ways Sarah incorporated this theme through design elements. Her opening page includes the suitcase images and the feeling of a global perspective. Notice how her sections include city images

FIGURE 3.2
Sarah's Design Elements
Source: Used with permission, Sarah
Milligan Weldon 2005.

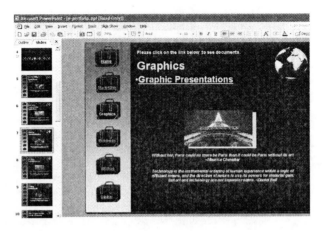

FIGURE 3.3
Sarah's Thematic Section Page
Source: Used with permission, Sarah
Milligan Weldon 2005.

that she chose to match each of her areas of specialization. Figure 3.3 shows the ways she incorporated these themes on one of her secondary pages. She carries the themes throughout and reinforces them through quotes on each page that refer to the particular city and to her skills as a technical communicator. The following quotations appear in her Paris/Graphics section:

> *Without her, Paris could no more be Paris than it could be Paris without its art.*   Maurice Chevalier

> *Technology is the instrumental ordering of human experience within a logic of efficient means, and the direction of nature to use its powers for material gain. But art and technology are not separate realms.*   Daniel Bell

Many of Sarah's other documents also create consistency across the portfolio. Through her work on theme and metaphor, she noticed patterns and the ways this travel theme was already (unconsciously) embedded in her artifacts. For example, her "Creative Resume" piece (Figure 3.4) (which was created long before her portfolio) shows her desire for adventure and her feeling about travel through the image of a passport.

Both Sarah and Miranda wanted their portfolios to express their personalities. Their designs are based on how they see themselves in terms of their personal traits and professional goals.

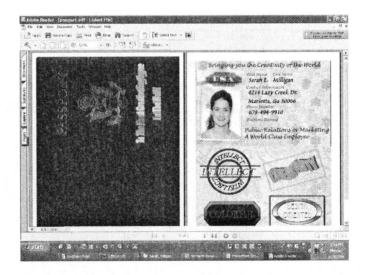

FIGURE 3.4
Sarah's Creative Resume—
A Passport
Source: Used with permission, Sarah
Milligan Weldon 2005.

Theme through personal interests

Some students, like Brian D., created their themes by connecting to outside interests that drove their choices of metaphor and design elements. Although he was a bit tenuous at first, Brian's theme ended up giving his portfolio a coherent feel and distinct personality. He says:

> Initially, I was resistant to the idea of incorporating a metaphor into the portfolio. I wanted my portfolio to have a clean, professional feel, and I felt that a metaphor would compromise that. I also wasn't sure about how to incorporate all of my interests with regard to technical writing. As it turns out, I was able to come up with a metaphor that I think works very well and incorporates some of my personality into my portfolio while at the same time keeping a clean, professional look. I used my creative resume of the tech writer marathon as a base for my metaphor and built off of that. I managed to incorporate the metaphor in a way that doesn't seem unprofessional, and I believe it actually adds a lot to my portfolio. And, if I'm lucky enough to get a job interview with someone who runs, we'll have something in common to discuss. That certainly can't hurt.

Notice Brian's sensitivity to the danger of compromising the quality and integrity of his work. There is always the problem, as Brian suggests, that the theme can distract attention from the portfolio rather than enhance it. It is important to consider the different people who will be viewing your portfolio and how they might interpret the theme. You do not want a theme that overshadows the contents or unintentionally excludes or offends particular audiences.

Brian has found a theme that draws on his passion for long distance running and recognizes the metaphorical connections between running, career advancement, and motivation, drawing on the jargon and symbols associated with the running culture. Brian notes that his metaphor might work on another level as well, attracting a particular community of runners should they get the chance to view his portfolio. Here is an image (Figure 3.5) from his home page in which he creates the metaphor of the "Tech Writer Marathon." He uses the image of the course map to show the span of his work and individual "mile markers" to indicate different sections and content. The theme is supported by energetic writing that explains the overall concept and section contents. Brian goes beyond the image to articulate and integrate the theme through reinforcing context statements. When carrying out a theme, it is important to find ways to create consistency throughout the portfolio in terms of design, graphics, and copy.

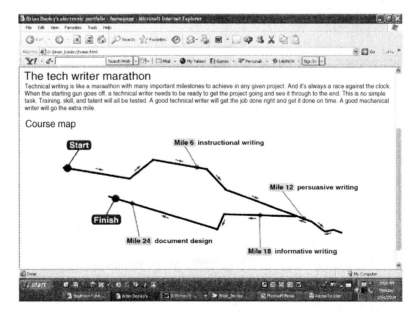

FIGURE 3.5
Brian's Home Page—Marathon
Source: Used with permission, Brian Dooley 2005.

FIGURE 3.6
Brian W.'s Race Flag Image
Source: Used with permission from Brian Wray.

Brian W, another student, is a race car enthusiast. He was looking for ways to incorporate his interests in multimedia and race car driving and decided on a theme that evoked the "racing look" in his portfolio. He explains:

> I prefer a racing, motor sports theme because of my interests in this field. I plan on using three colors derived from BMW's Motor Sports division, which include light blue, dark blue, and red. I will also use a checkered flag theme throughout the portfolio. Each section will have a different color for its theme.

This was a natural fit for Brian, as much of his material also focused on race car driving as part of his content.

Figure 3.6 was part of Brian's research to determine his theme. Here we see that he was intrigued by the images of the blowing checkered race flag, which he later incorporated into his background design through an electronic watermark. He also drew on other images, such as the race track, font styles, and car designs. This made sense because many of Brian's documents also harkened back to his love of the sport—a connection he did not realize until he reflected on his work. For example, notice that in Figure 3.7, an artifact from his portfolio, he uses the subject and images of cars to demonstrate his skills in the graphic design of this chart.

Both of these examples show ways you might arrive at a theme through personal interests. However, when choosing an interest, pay attention to the messages you are communicating. Although interests create convenient metaphorical connections, you sometimes risk having too narrow a focus or the chance that an employer might confuse your theme with your professional interests or skills. For example, one of our

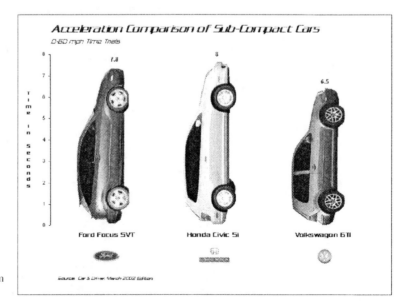

FIGURE 3.7
Brian W.'s Race Car Chart
Source: Used with permission from Brian Wray.

students started out with the image of a cell phone to indicate communication but later considered it confusing to associate himself so closely with the wireless industry since working in that industry was not one of his professional goals. As noted previously, metaphors are deeply and subconsciously ingrained in our culture. Therefore, it is important to avoid symbols that are overly saturated or ones that convey preformed cultural messages that you might not be able to modify.

Conceptual theme

Another way to pull a theme into your portfolios is to create a conceptual metaphor. Here are some examples from two other students, Amy and Norma, who used philosophical concepts to create themes. Amy drew on the concept of a jigsaw puzzle and found graphic elements to match:

> I want to show that I am a critical thinker, yet creative and interested in many things. The metaphor that is coming to mind is something like a jigsaw puzzle. I would have the pieces of the puzzle represent different kinds of things I can do; yet when all the pieces are put together, you have a better idea of the big picture.

Amy chose a symbol that represents her image of herself as a "critical thinker" who is "interested in many things." The puzzle pieces allowed her to show her separate areas of interest (which work their way into sections), and the completed jigsaw puzzle represents the ways these pieces come together to form a highly marketable employee. Here is an excerpt from Amy's introduction, in which she explains the metaphor:

> There are so many different things that make up who I am. You could ask people who know me what things make up Amy, and they would tell you that I am like a puzzle. There are some people who never get to see the whole picture. Some pieces I choose to hide. Some of the pieces that people see in a social setting simply do not come out in an academic setting.

She extends the metaphor by detailing the assembly of these pieces she brings together:

> This portfolio helps to bring these pieces together. I have samples from the things that interest me and that help to define who I am.

She carries out this explanation in her graphic section dividers in Figure 3.8. She uses the puzzle imagery and differentiation through color (for different sections) to extend the theme throughout the portfolio.

Norma also developed a conceptual theme growing from the fact that she had experienced a lot of change in her career and personal development. She wanted her portfolio to demonstrate this evolution along with her awareness of these changes and chose a butterfly—an image that she had been attracted to since childhood—to create a theme linking her portfolio:

> I was afraid my butterfly metaphor might come off as too childish or not professional enough. But I realized that the metaphor was true to me. It spoke to my development from a computer science student into a technical writer.

Like Brian D., Norma was a bit unsure of her choice, but she pulled it off with a sense of professionalism and creativity. She used the metamorphosis of the butterfly to show her own changes as a writer and communicator, drawing on the brilliant, vivid hues of butterfly's wings to create a color scheme, design elements, and texture. Figure 3.9 shows one of the butterfly images she used to inspire the design elements of her portfolio.

Norma's paper portfolio is full of these rich colors, and the background color of each section is coordinated with textured, multicolored papers and materials. In her electronic portfolio, she uses different colors and

FIGURE 3.8
Amy's Puzzle Design
Source: Used with permission, Amy
Grau 2005.

FIGURE 3.9
Norma's Butterfly Image
Source: Norma O. Gonzalez.

types of butterflies for each of the section links. Her introductory sections for both the paper and electronic portfolios include her own story of change as she developed her skills as a communicator. Both portfolios also feature quotations that reinforce and highlight the themes of metamorphosis and change. For example, the opening of her section on professional communication expresses her desire to succeed and her willingness to do whatever it takes to get a job done. She quotes author Trina Paulus (1972):

> "How does one become a butterfly?" she asked pensively. You must want to fly so much that you are willing to give up being a caterpillar.

Figure 3.10 shows one of the sections from Norma's electronic portfolio. You can see how her theme is applied. Obviously, this kind of symbolism would not work for everyone, but it meant something to Norma, who closely connected her identity to the image of the butterfly.

Michel, on the other hand, wanted a more conservative look since she was trying to enter the field of medical communication with a government organization. She focused on the medical community to showcase her medical writing and create a portfolio to meet the conventions and expectations of that field:

> I put all my medical stuff up front. It's the first and longest section of my portfolio. I think most of my best work is in there, too. Every piece in that section exemplifies what I think medical writing should be—easy to understand and fact-driven.

Visually, her portfolio has a classical look that fits with the scientific feeling of her field but at the same time shows a sense of style:

> The idea appeared fully formed in my imagination—burgundy leather, cream pages, old-fashioned script titles, and all.

Most of the artifacts Michel chose to include helped to create the focus on medical writing. However, she also included additional sections so that she can present herself as a generalist should her ideal job not work out.

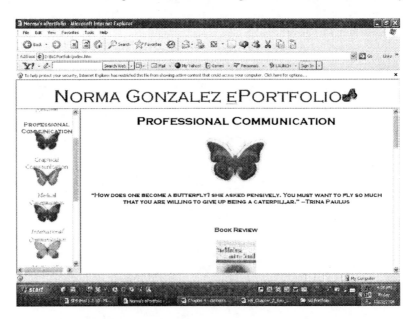

FIGURE 3.10
Norma's Home Page
Source: Norma O. Gonzalez.

(Michel did get the job she desired, and she credits her portfolio for creating the opportunity.) Notice how, in Figure 3.11, her table of contents represents her goals and focus.

Michel offers a range of medical writings including three different pieces—a journal article, a press release, and a brochure—on the same topic: tularemia. By packaging her medical writings in this fashion, Michel shows that she can present similar subject matter to multiple audiences using a variety of document designs, a useful skill in the medical writing profession. Michel also has decided to include some of her best creative writings but places them at the end of her portfolio because these pieces do not relate directly to her medical writing goal.

Michel's Table of Contents

Resume

Medical Writings

    Research paper: "Child Life or Wild Life: Should DDT Be Used to Control Malaria?"

    Feature Article: "Worse Than Worthless—The Over-Prescription of Pediatric Antibiotics"

    Brochure: "Living with Stress"

    Journal Abstract: "Tularemia as a Biological Weapon"

    Press release: "Third Confirmed Tularemia Case Fuels Rumors of Bioterrorism"

    Feature article: "Mystery at Martha's Vineyard"

    Brochure: "Tularemia"

Business Writings

    Press release: "ING America's Announces New Head of Diversity and Community Relations"

    Sales letter: "So Close to Real"

    Book Review: "Betrayal of Trust: The Collapse of Global Public Health"

    Research paper: "Marketing the Macabre: The Faces of Shock Advertising"

Graphics and Multimedia

    Creative resume: "Think All Chicks Are the Same?"

    Logo design: "M. Stevens, Science Writer"

    Excel graphics: "Mother Knows Best"

    PowerPoint Slide Revisions: "Before and After"

    One-page document: "Mata Hari: Lady of Intrigue"

    Multimedia presentation: "Palomino Ranch"

Creative Writings

    Short story: "Cardboard Caverns"

    Short story: "Family Endures"

    Poem: "My Father's Work"

FIGURE 3.11
Michel's Table of Contents
Source: Michel Alexander.

Jarmon, a technical communicator with an electrical engineering minor, decided to arrange the contents of his portfolio to emphasize his graphics and media skills rather than the writing skills Michel chose to highlight. The table of contents for his electronic portfolio, shown in Figure 3.12, opens with a section on graphic design, which is supported by a later section on multimedia design. Because Jarmon has a minor specialization, he also chose to include a couple of engineering documents, specifically a project report and a lab report, in his technical documentation section. Jarmon's portfolio emphasizes the strong connection between his technical communication major and his electrical engineering minor (Figure 3.12). He reinforces his specialization area through design elements such as circuitry, plugs, wires, and connectors—conceptual images associated with the field of electrical engineering.

Trina, on the other hand, wanted to show her ability "to speak out without fear." She wanted employers to know about her courage in addressing controversial subjects and her willingness to take on leadership roles. In the process of brainstorming, Trina came up with a "Mrs. Potato Head photographed with various large mouths to display that quality in myself." Although she eventually moved away from this image because "it seemed less professional" and too comical (and is a copyright protected image), it made her reflect upon

---

Jarmon's Table of Contents

Resume

Graphic Design

    Gray Consulting Business Letterhead

    Gray Consulting Business Card

    Gray Consulting Business Envelope

Document Design

    Michael Jordan: The Greatest

    Jordan Surveys

    Career Services Pamphlet

Technical Documentation

    Traffic Light Design

    Slotted Line Measurements

    RoboHelp Training Manual

Multimedia Design

    RoboHelp Software

    How to Maintain Your Car

    When Technology Fails

Organizational Management

    NSBE Black History Production Flyer

    Faculty-Student Basketball Game

    Health Field ID Cards

FIGURE 3.12
Jarmon's Table of Contents
Source: Used with permission, Jarmon Gray 2005.

choices about how to represent herself. This idea morphed into her final theme, in which she used quotations from and images of courageous women she respected to head each section and provide a link to her contents. Both of these ideas were connected to her desire to show that she is not afraid to speak out. It is important to recognize that it often takes persistence to find a theme and a metaphor that are both appropriate and effective. Trina's willingness to be open during her brainstorming phase allowed her to think creatively and eventually refine her ideas into a strong theme with a powerful message.

All of these students used conceptual themes to create their portfolios, whether related to their professional goals, areas of specialization, or philosophical beliefs.

Theme through symbols

A symbol can also provide a thematic and graphic element to bring consistency to your portfolio. For example, Nanette wrapped her theme around a symbol. In one of her earlier works, she completed a creative graphic piece in which she incorporated street signs and the information highway. In her opening page, she presents the overall idea of "All Roads Lead to Technical Communication," creating a graphic of an old-fashioned street signpost in which the signs pointed in different directions. Each sign represented a section or skill that she had developed. Figure 3.13 shows a page from one of Nanette's opening sections in which she continues to use the sign as a link to different sections.

The examples described in this section show just some of the ways you might arrive at a metaphor or theme for your portfolio. We know that this is one of the most difficult parts of the process to conceptualize, but the results are worth it. Time and time again, we have seen that the strongest portfolios were the ones demonstrating a consistent look and feel. Although this can be achieved through design alone, we feel that the marriage of design and content through theme is the best way to ensure consistency. Your theme or metaphor will also help you to stand out as an individual for employers who are often inundated with cookie-cutter portfolios designed using generic templates. Work to develop your own metaphors and thematic connections to give your portfolios a sense of continuity and individuality.

Use the examples and the scenarios described in this chapter to complete Exercises 3.4 and 3.5, which will help you think about ways to incorporate theme and metaphor into your own portfolio. These exercises will help you to see your work in new ways as you notice recurring patterns and images. They will also help you to communicate your purposes more actively to your audience as you reshape your rhetorical choices around new subjects and images. Finally, you will be able to draw upon these themes and metaphors as you create your portfolio proposal, table of contents, and presentation.

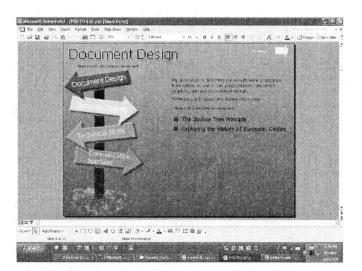

FIGURE 3.13
Nanette's Section Graphics
Source: Used with permission, Nanette Packman 2005.

EXERCISE 3.4 THINKING ABOUT METAPHOR AND THEME

In this exercise, we take you through heuristic activities that should help you explore possibilities for a metaphor or theme. Take several minutes to respond in writing to each of them. Start by looking through your work and making notes as you read.

1. After looking through your work, list and describe any themes or patterns you see emerging from your documents, graphics, or multimedia pieces.
2. List and describe the images that stand out.
3. List and describe the concepts that stand out.
4. Describe any personal interests that emerge from your artifacts.
5. Create a list of words that describe your work ("bold," "understated," "professional," "creative," "technical," etc.).
6. List six words that describe your personality.
7. List six words that describe your professional goals.
8. List six words that describe your skills.
9. Describe your collection of work as a color or group of colors.
10. Complete this sentence: My work is like a _____.
11. List six objects or symbols that might represent your work.
12. List six images that you find interesting or engaging.

EXERCISE 3.5 INTERNET RESEARCH: EXPLORING THEME THROUGH IMAGES

Once you have completed your metaphor/theme heuristic, read it and use the ideas and terms you generated to create searches on the Internet. Note the examples described by our students in this chapter (Sarah: travel and entertainment; Michel: medical; Miranda: Art Deco; Nanette: signs; Brian: running). Follow this research path as you key in different search terms and ideas. Do the same thing with an Internet image search in which you use the same terms to search image files for brainstorming purposes. Follow where this search takes you, making special note of new terms and images that emerge. Take notes along the way, including citation information.

CREATING A PORTFOLIO DESIGN PROPOSAL AND TABLE OF CONTENTS

All of the work in the preceding chapters should lead you to the point where you are ready to create the plan and overall design for your portfolios. Recognize that, like any act of composition and revision, your plan is subject to change and growth. In this book, we present revision as a recursive process in which you can revisit your ideas at any time. However, at this point in the process, it is important that you have a general point from which to start.

Reread the exercises you have completed. As suggested throughout this chapter, you will make important decisions about design, organization, and structure. Start by composing a statement in which you begin to structure the contents of your portfolio. As you will see in many of the student statements that follow, it is time to start ranking your priorities and thinking about order and presentation. Here are some ideas generated by our students regarding their decisions about organization, design, and content at this point:

※ Sarah: I am organizing by a natural order, with the flashier multimedia at the front and the writings at the back. I wish to add a section at the back for photography and original recipes (for a possible job or internship with a food show of some sort).

* Miranda: I can show original drafts. I'll focus on the process by showing planning documents, and I will also show multiple drafts that lead to a finished piece. That will give me the opportunity to talk about my choices.

* Michel: I began to categorize all of my documents into communication skills I felt were important to highlight. For example, I decided to include medical communication, professional communication, and even creative communication. At first, I was not sure about including creative communication. But that too is part of who I am. I wanted to showcase my technical and creative sides at the same time.

* Norma: I want to include original and revised versions of some of my works. I want to include ten to fifteen pieces to show my work. I want to have a metaphor for my portfolio. I'd like to have butterflies as my metaphor and theme throughout my portfolio. I have always loved butterflies and feel that they would represent my growth and development as a technical writer.

* Brian D.: I have four sections representing the different genres of tech writing. The four sections show my main interests and skills, while at the same time show that I can do a wide variety of different things. For some people this may seem like a lack of focus, but for me, I wanted to show my ability to do lots of different things—especially as an entry-level writer.

* Amy: I will have a section displaying my critical thinking and research skills that will be comprised of at least two essays. I will have another section displaying the work I have done for various on-campus groups that I'm involved with that contains materials such as brochures, manuals, and Web sites. I will have a section that covers the skills I developed in my Technical Training class, and I will also display some of my editing work. As a lighter note, I will display some of the journalistic pieces I have written. I think my strengths are best displayed through my writing (journalism, technical training, etc.) and my Web work.

All of these students are thinking about important rhetorical choices that involve organization, style, theme, and content. You will see samples of their final design proposals and table of contents throughout (and at the end of) this chapter that should help you think about some of your decisions. Before creating your table of contents, reflect upon the questions in Exercise 3.6 to help you articulate the organization and goals of your portfolio. Although you have done some of this initial thinking in earlier chapters, this exercise asks you to reconsider all you have learned in light of the design of your portfolios.

EXERCISE 3.6 PORTFOLIO HEURISTIC

Establishing Criteria for Portfolios

Answer the following questions regarding your portfolio. Based on these responses, in a short writing activity, explore possible directions for your class portfolios (use a separate sheet of paper).

1. What are the criteria for good writing/technical skills established in this context/class/profession?
2. What skills do I hope to present in my portfolio?
3. What should be included in the portfolio? What will best demonstrate my work as a writer/professional? What gaps do I see at this time?
4. What audiences and purposes should my portfolio serve? How will my presentation choices work for multiple audiences and purposes?
5. What organizational and style issues should I consider in presenting my portfolio?
6. How will my portfolio reflect the process of writing, including invention, drafting, revision, and editing?
7. Will I have a theme or metaphor for my portfolio? What themes or metaphors best represent my work and philosophy as a whole?
8. How will I visually/graphically represent my portfolio?
9. How will my electronic portfolio differ from my paper version?
10. How will I present my portfolio to others (include materials and settings)?

After you have reflected upon your goals, presentation style, theme, and contents, it is helpful to develop a formal proposal and table of contents that will act as your guide throughout the rest of the process. The exercises at the end of the chapter will help you to come up with your own versions of these documents that will serve as guides.

We have included a few sample tables of contents (Figures 3.14 to 3.18) in Appendix A at the end of this chapter to give you an idea of how different people organize their work.

As Miranda indicated in the chapter-opening quotation, it is important to have a plan to keep you organized and consistent. All the student samples displayed resulted from much reflection and writing about their goals, audiences, and purposes. As these samples show, the students used this work to come up with focused proposals and tables of contents to organize their artifacts and connect them via themes and purposes. If you have the opportunity, have your proposal and table of contents critiqued by an informed audience. If you are in a classroom setting, your classmates might participate in a response session at this phase of your portfolio design. If you are in a professional setting, you might get productive feedback from a colleague or mentor. The more you expose your developing ideas to a real audience, the more effective your portfolio will be. This will give you a chance to revise it to achieve a winning presentation. For example, Miranda created a PowerPoint presentation that she used to showcase her approach, style, and theme. We have included it (Figure 3.19) at the end of the chapter to show how you might present your ideas to an audience for feedback. Although some of these drafts changed before their final presentation, they provided the students with a comfortable structure to return to throughout the process of constructing their portfolios.

SUMMARY

Chapter 3 introduces the concepts necessary to begin the process of designing your portfolio. It discusses the ways you might represent your skills and goals through careful selection and organization of your artifacts, using the concept that portfolios are representative rather than comprehensive. By the time you complete the following exercises, you should be ready to start revising your artifacts and constructing your portfolios. These exercises will guide you through the process of developing your proposal, table of contents, and feedback presentation.

ASSIGNMENTS

Assignment 1: *Portfolio Design Proposal*

Use the heuristics and analytical work completed in the earlier chapters. For this assignment, you will plan your portfolio by addressing the following parts:

PROPOSAL

Write a short abstract in which you describe the purposes of and audiences for your portfolio. Discuss organization, style considerations, and visual strategies. Propose an overall design and approach. Discuss a theme or metaphor that you will use throughout your portfolio. Pull ideas from the Portfolio Heuristic (Exercise 3.6) to shape your portfolio. Write a short introduction for each portfolio section that you will eventually revise and include in the portfolio. These introductions should indicate how the sections demonstrate your skills and strengths in a particular area or time period.

VISUAL ELEMENTS

Incorporate any graphic elements and fonts/styles you are considering, as the introductions and table of contents should represent the visual style of the whole portfolio. Discuss colors and style.

MATERIALS

List the materials needed to assemble your portfolio (notebooks, divider pages, labels, photo paper, etc.). Be as specific as possible, and try to list particular types and brands.

QUESTIONS/GAPS

At the end of your writing, discuss any lingering questions or gaps that you see in your portfolio. This is the place to list projects that you need to generate or revise. It will also give you a chance to get feedback from a live audience.

Assignment 2: *Table of Contents*

Prepare a table of contents listing all of the documents in the portfolio. Title each of your artifacts, and divide the table of contents into sections and title each section. Be as specific as possible (e.g., don't just say "a graphics piece"; instead, label it as a "logo and letterhead" or provide a descriptive title such as "Multimedia Presentation: The Wild West"). Although this document should demonstrate the actual order of your artifacts, do not include page numbers at this time. The table of contents should reflect your decisions about color, design elements, and document design (including font choices). This will help you to visualize it as an integral part of your portfolio.

Assignment 3: *PowerPoint Presentation*

If given the opportunity, have your proposal and table of contents critiqued by an informed audience. If you are in a classroom setting, your classmates might participate in a response session at this phase in your process of portfolio design. In a professional setting, you might also get productive feedback from a colleague or mentor. The more you expose your developing ideas to a real audience, the more effective your portfolio will be.

PREPARE FOR PRESENTATION

In a short PowerPoint presentation to a small group or a single colleague, share your ideas for the plans and goals of your portfolio. This will be followed by a short workshop/critique session in which your classmates/colleagues will give you feedback for revision. Your presentation should address the following (see Figure 3.19):

* Title
* Purpose and audience
* Theme or metaphor (history, connections, description of design elements)
* Section layout or hierarchy (overall structure—table of contents)
* Images and icons (include actual images)
* Fonts and design elements
* Style and layout
* Navigational structure

Prepare a short narrative presentation to go with your PowerPoint slides.

Assignment 4: *Project Management Timeline and Working Portfolio*

Create a projected timeline with both short-term and long-term goals for managing your portfolio project. This will help you consider layout, audience feedback, user testing, final revision, and so on. Include all the details—both large and small—that must be completed in order to finish your portfolios. Create a project sheet that breaks down the process into manageable, tangible units.

As you plan, build and revise your work, you should keep all documents in a *working* portfolio—your portfolio in progress. This will help you to organize and preserve your work. Design a three-ring binder in which you include page protectors and dividers that reflect the tentative sections of your portfolio. As you revise the artifacts, place them in this working portfolio. Along with providing protection for your documents, this working portfolio allows you the flexibility to change the order and emphasis as you revise the global structure of your portfolios. You will need the following materials:

* Three-ring binder
* Page protector sheets
* Tab dividers

REFERENCES

Lakoff, George and Mark Johnson. *Metaphors We Live By*. Chicago: University of Chicago Press, 1980.

Paulus, Trina. *Hope for the Flowers*. New York: Paulist Press, 1972.

APPENDIX A

Figures 3.14 to 3.18 present the tables of contents of five of our students. Figure 3.19 shows Miranda's PowerPoint presentation.

Sarah's Table of Contents

Introduction
Resume
Creative Resume

Marketing
Document Design
 Homer Simpson
 Harry Potter
 Invitation
Presentations
 Destination Wedding
 Event Solutions Website

Graphics
 Country Graph
 Birth Map

Business
 Logo Design
 Departmental Newsletter
 Delta Airlines Letter

Writing
 CCC Brochure
 Communications Between the Sexes

Global Communications
 South Korean Business Practices
 North West African Presentation
 Eiffel Tower Short Paper

FIGURE 3.14
Sarah's Table of Contents
Source: Used with permission, Sarah Milligan Weldon 2005.

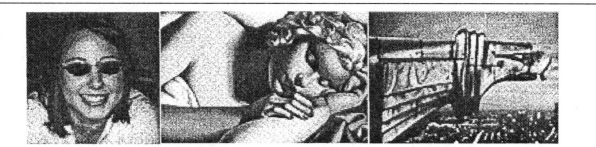

Miranda Bennett
Portfolio Outline
Spring 2004

Multimedia

Web sites

<u>The Kai Review Web site</u>

The Kai Review showcases my ability to **create web graphics** and use **CSS for both style and positioning.** I designed this Web site as part of a group project in Web Design. *The Kai Review* was intended to become a printed literature magazine at SPSU. However, the editorial board's interest waned after the class. In order to give a clear picture of the intended Web zine, I have used public domain content to fill in areas where content was either insufficient or missing.

Tools Used:
Pad+2
Adobe Photoshop
Microsoft Word

<u>Social Constructionism Web site</u>
The Web site showcases my ability to **create a Web site quickly.** I designed this Web site for a group project in the Research in TCOM class (spring, 2004). Using Adobe GoLive 5.0,1 was able to create this Web site in approximately **two hours.** The Web site served as a teaching tool on social constructivism for the class. In addition to designing the Web site, I also wrote the introduction on the home page of the site.

Tools Used:
Adobe GoLive 5.0
Adobe Photoshop
Microsoft Word

Help Files

<u>Pad+2 Help File</u>
I created this help file for my Online Communications class. Although the help file was never used with the application, it allowed me to segue into a position as a consultant for the company, Howell Developments.

Tools Used:
Macromedia RoboHelp
Adobe Photoshop
Microsoft Word

Fig. 3.15 continued ....

RSS Composer Help file
I created this help file under a very tight deadline. Howell Developments had recently become involved in the syndication community. They needed a help file within three days to meet their release date of February 10, 2004. The help file I created was packaged with the application. Howell Developments has since put the pre-release for a more comprehensive syndication application on the market. On this help file, I delivered the HTML files. The company assembled the files.

Tools Used:
Microsoft Word

Design

Publications

Old Sting compared to previous layout
As the Editor-in-Chief of *The Sting* at SPSU, I was able to experiment with the layout and design of the paper. I am most proud of the entry points I created by incorporating thumbnails in the left column on the front page. I created more, and stronger, visual entry points throughout the newspaper by using larger images and by grouping images to pull in the reader. I also decreased white space.

ELS Voice layout
The ELS voice is another example of my ability to work under pressure. I laid this entire publication out for the ELS students at SPSU in just under three hours, at the request of a fellow student. I was happy to help out, but also enjoyed the creative freedom I was given over the publication.

GA Map of Life – Arts 3000
I created the GA Map of Life in my Arts 3000 class. It shows my progression from childhood to college and from south Georgia to north Georgia.

Tools Used:
Adobe Photoshop

Blueberry Festival Advertisement
I created the Blueberry Festival Advertisement while working for *The Douglas Enterprise*. The advertisement ran in both *The Douglas Enterprise* and the weekly shopper called The Bonus.

Tools Used:
MultiAd Creator
Adobe Photoshop

Fliers

Orientation Flier
The Sting Flier is a marketing piece to encourage incoming freshmen at SPSU to join the student newspaper staff. It was included in the 800 folders given out for orientation during the fall semester of 2003.

SGA Flier – 3 together
I created the three fliers during my campaign for the Student Government Association (SGA) Presidency at SPSU. I was one of three candidates, who also included the incumbent and the current vice president. My fliers answered questions from the incumbent concerning my qualifications and also spoke to current policies of the SGA. I lost the campaign. However, I received a significant number of votes and a great deal of experience communicating under pressure and on a large scale.

Tools Used:
Adobe PageMaker
Adobe Photoshop
Microsoft Word

continued ....

Graphics

CAB Casino Night
I created this advertisement for the Campus Activities Board's Casino Night and used it in the college newspaper, *The Sting,* of which I was Editor-in-Chief.

Tools Used:
Adobe Illustrator
Adobe Photoshop

SOAR
I created the SOAR Concept page for the brochure mailed to seminar participants when I worked as an intern for a small Altlanta-based actuary firm. I also improved the logo used on the Web site to make it pop off the page.

Tools Used:
Adobe Illustrator
Adobe Photoshop

Pixar Bar Chart
I created the Pixar Bar Chart for Foundations of Graphics to show an alternative way to present information in a bar chart format. Rather than use bars, I used characters from the movies each bar represented.

Tools Used:
Microsoft Word
Microsoft Excel
Adobe Photoshop

Ethics Table
I made the Ethics Table for a group presentation in my Foundations of TCOM. The table compares student responses (on the left) to responses from TCOM professionals (on the right) in regard to how to ethically use and represent data in different formats. The graphics of the student and man were taken from the Internet.

Tools Used:
Adobe Photoshop

Writing

Technical Writing

Researchers Seeking Smaller Chips
I created this article for my Production Processes class, part of the Industrial Engineering Technology Department. The article is about current microchip manufacturing processes and two new processes for chip manufacturing.

Tools Used:
Adobe PageMaker
Adobe Photoshop
Microsoft Word

Brain Science Book Review
I wrote this book review of *Why God Won't Go Away* for my Science Writing class.

Tools Used:
Adobe PageMaker
Adobe Photoshop
Microsoft Word

Adobe PageMaker Training Manual
As an assignment for my Technical Training class, I wrote a training manual to be used in a beginner's workshop for learning PageMaker 6.5. This training manual included pre-tests, post-tests, criteria checklists to measure progress, a training strategy, and, of course, the training exercises. The manual totals 40 pages. I have included only 5 pages as a representative sample.

continued ....

Tools Used:
Adobe PageMaker
Adobe Photoshop
Microsoft Word

<u>PX Gallery Information Maps</u>
As an assignment for my Manuals class, I wrote these information maps about features of PX Gallery Wizard (software from Howell Developments used for generating online photo galleries).

Tools Used:
Microsoft Word

FIGURE 3.15
Miranda's Table of Contents
Source: Used with permission, Miranda Bennett 2005.

Michel's Table of Contents

Resume

Table of Contents

Medical Writings

Research Paper
Child Life or Wild Life: Should DDT be used to Control Malaria?
Feature Article
Worse Than Worthless—The Over-Prescription of Pediatric Antibiotics
Brochure
Living with Stress
Journal Abstract
Tularemia as a Biological Weapon
Press Release
Third Confirmed Tularemia Case Fuels Rumors of Bioterrorism
Feature Article
Mystery at Martha's Vineyard
Brochure
Tularemia

Business Writings

Press Release
ING America's Announces New Head of Diversity and Community Relations
Sales Letter
So Close to Real
Book Review
Betrayal of Trust: The Collapse of Global Public Health
Research Paper
Marketing the Macabre: The Faces of Shock Advertising

FIGURE 3.16
Michel's Table of Contents
Source: Michel Alexander.

Norma's Table of Contents

Proposal

Professional Communication

- *The Medusa and the Snail* – Book Review
- *Finding a Job in TCOM* – Formal Memo Report
- *Orion Television Company* – Negative Adjustment Business Letter
- *Norma's Literacy Journey* – Radical Revision Writing and Editing Assignment

Graphical Communication

- Celebrating Elvis – 2 Page Layout
- Girl's Weekend Celebration – Flyer
- Norma's Soundtrack CD – Creative Resume
- Business Card

Medical Communication

- **Dissociative Identity Disorder (DID)**
  - *Coping by Forgetting* – Informative Article
  - *Government for Mental Health Organization* – Press Release
  - *About Dissociative Identity Disorder* – Informational Brochure
  - *Introduction to DID* – PowerPoint Presentation

International Communication

- Literatura Hispanica (Translation: English and Spanish versions)
- Ambassador Country Report: *South Africa*
- Ambassador Country Presentation: *South Africa*

Multimedia Communication

- Individual Presentation: *Journey of a Monarch Butterfly*
- Group Collaboration: *Peachtree Center Presentation*

Creative Communication

- *I-Ching and Bamboo* (Painting – Acrylic on Canvas)
- *Lion Awaiting* (Drawing – Charcoal on Paper)
- *True Love* (Graphic – Created in Adobe Illustrator)

FIGURE 3.17
Norma's Table of Contents
Source: Norma O. Gonzalez.

Brian's Table of Contents

Multimedia

- EPSS Professional Video Presentation
- Jeep Grand Cherokee PowerPoint Presentation
- Maxim Fashion Magazine Video Commercial
- Personal Professional Web site
- "The Good Vodka" Short Film

Writings

- Social Engineering Rhetorical Analysis Paper
- Counterterrorism Responsiveness Paper
- Once and for All Pest Control Sales Letter
- Morpheus Software Manual

Graphic Design

- A Supercharged Grand Cherokee? Two-Page Spread
- The First 1300 Miles Magazine Page
- Creative Resume
- Acceleration Comparison of Sub-Compact Cars Bar Graph

About Me

- Racing
- Scuba Diving
- Martial Arts
- SETI@Home

FIGURE 3.18
Brian W.'s Table of Contents
Source: Used with permission from Brian Wray 2005.

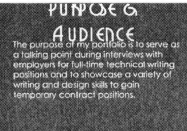

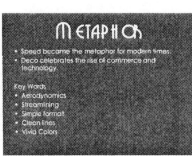

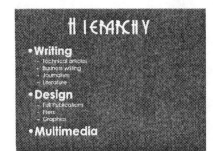

FIGURE 3.19
Miranda's PowerPoint Presentation
Source: Used with permission, Miranda Bennett 2005.

Fig. 3.19 continued ....

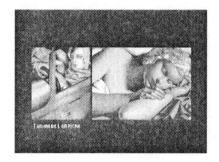

FIGURE 3.19   *Continued from page 57*
Miranda's PowerPoint Presentation
Source: Used with permission, Miranda Bennett 2005.

4 Revising for Portfolio Quality

To revise, writers must compare versions, consider alternate methods of development and organization, assess the quality of their communication, and play, often with style options. (Bishop 2004, vi)

Once the document design bug bites you, you never look at documents in the same way again. Sarah

INTRODUCTION

This chapter helps you to understand the methods of revising your documents and artifacts to achieve portfolio quality, guiding you through the rhetorical and design decisions that will enable you to show a variety of skills for new audiences and purposes. You will also find suggestions for moving from a collection of documents to a comprehensive portfolio concept. The chapter discusses the importance of both local and global revision strategies, including the creation of context and thematic organization. Chapter 4 covers the following topics:

- Revision: a recursive process
- Revising locally
- Revising for new audiences and purposes
- Revising for content
- Amplifying: representing multiple skills within a single document
- Editing
- Revising globally
- More than a container: contextualizing artifacts
- Overall structure and context

REVISION: A RECURSIVE PROCESS

Writing theorists generally agree that revision is a recursive process involving more than a lockstep approach that moves from one draft to the next. Revision evolves along with you and is continually affected by many variables. As a language user, you are continually revising your words and ideas in relation to your audiences, purposes, and contexts. At any point in the processes of writing, you might go back to address invention, organization, and style issues.

As the chapter-opening quotations indicate, revision is a complicated ongoing series of processes. As a writer, communicator, and designer, you have encountered many forms of revision with varying levels of intensity. This next phase in the development of your portfolio asks you re-see and revise.

Once you have a sense of direction and a general plan, you must focus on the ways your artifacts come together to form a complete body of work. Carefully consider the pieces to include in your portfolio, as well as the messages you wish to communicate to others. The creation of your portfolio is an intense act of revision in which you re-see your ideas, texts, and images in a new rhetorical situation. In previous chapters, you analyzed the rhetorical situation—your purposes, audiences, content, and context. This chapter asks you to put that knowledge to use as you revise.

For the purposes of this book we separate the terms "revision" and "editing." We draw upon the distinctions made by Wendy Bishop in *Acts of Revision* (2004), who suggests that **revision** is the "practice of making meaningful changes in texts at the word, sentence, paragraph and full text level by adding deleting, substituting, and rethinking their work entirely" (vi), whereas **editing** involves "assuring that the text has consistency and, at times, that it conforms to a 'house style': the requirements of a class or publisher" (vi) (and to this list we would add "client"). Editing involves careful examination and correction of surface errors such as spelling, grammar, or mechanical mistakes. It is an important part of the process that is best saved for the end. Carefully edited (or, for that matter, sloppily edited) materials definitely reflect your skills as a communicator. When we speak of revising, we refer to the conceptual and stylistic changes that you might make as a writer or designer in developing your portfolio.

In this chapter, we focus on three levels of portfolio revision, relying on the categories suggested by response theorists Richard Straub and Ron Lunsford (1995) in their model for analyzing texts: local issues, global issues, and context issues. For them, local issues "deal with microscopic areas in student texts" such as structure, wording, and correctness. Global issues "deal with macroscopic issues in the text or global matters" such as ideas, organization, and overall theme. The last category they describe deals with context elements that "go beyond the immediate concerns of the text and deal with the contexts that inform the piece of writing" such as rhetorical context, physical or social setting, or industry standards (159). We draw on these distinctions to discuss revision strategies for your portfolios. For our purposes, local revisions include work on the individual documents, images, and other products within the portfolio—the artifacts or microscopic areas in your texts. Global revisions refer to the ways you see the collection as a whole in terms of theme, structure, and overall communication of ideas, theme, and professional identity. Context revisions refer to the ways you consider your rhetorical situation and factors such as industry standards and usability. Table 4.1 details these types of revisions as they apply to your portfolios.

REVISING LOCALLY

Obviously, you want the artifacts in your portfolio to look their best. Many of them were created earlier in time. Active materials such as brochures, videos, Web sites, or published work should probably remain as they were originally presented, particularly if they were used publicly or in the context of a job or contract work. If you have this professional work, it is obviously best to include as much of it as possible (make sure that you obtain permission). However, many of you are students or in entry-level jobs in which you generated material while learning. This process asks you to reconsider your work and revise it for a new context. In our experience, at this point in the process, people often read their collection of work and say, "I can't believe I wrote this" or "If I only knew then what I know now." This is perfectly natural, especially for students of

TABLE 4.1
Types of Revision

Type of Revision	Focus	Elements
Local	Microscopic: documents and individual texts	Content, audience, visual and design elements, editing
Global	Macroscopic: the whole portfolio	Organization, theme, sections, visual consistency, readability
Context	Contextual factors: influential outside factors	Rhetorical situation, industry standards (including skills represented), and context statements (introduction, document, and context)

communication. We believe that life and experience are acts of revision. So when you face your documents again, hopefully you will be able to see them through the critical eye of your accumulated knowledge and experience. Or, as one of our students, Joy, says, "Portfolio quality means combining what you have learned throughout your education and applying it to all of your documents."

REVISING FOR NEW AUDIENCES AND PURPOSES

As we discussed earlier, you are reshaping your documents for new audiences and purposes. Many of the materials you use might have been appropriate for the audience they were originally created for but do not reflect a professional image. The element of audience requires that you know the conventions of your field and consider your new readers.

Note Sarah's responses as she looks back on her work in terms of revision:

> Reflecting back on my work, I saw serious changes that needed to be made. I couldn't believe some of the grades that I received on items that I would never hand in now (and a good measure of a learning curve). There are three pieces in particular that have changed dramatically. First, my H.P. piece. Originally, this piece was in black and white with okay graphics, at best. Now it is in full color, and a true testament of the kind of work that I can do given the tools.

Sarah comments, in particular, on the ways her graphic abilities have improved throughout her degree program. She refers to the tools she has mastered and thinks about the ways they can help demonstrate these skills through revision. She also thinks about revising her work in terms of what she now knows about graphic design:

> I really liked my first logo design when it was originally done in a graphics class a year ago; however, when I pulled it out to place it in the portfolio, I was shocked. ICK! I revised my original design and made it clearer, with better effects. I redesigned the product for my portfolio, and now I know that it looks much more professional and appropriate.

She takes an assignment that was written for a classroom audience and tries to see it through the eyes of a potential employer. She understands that her professional identity is moving from that of a student to a working professional, with higher standards and a stronger sense of responsibility that goes beyond a classroom grade. She strives for clarity and professional appeal.

Think about the differences between the audience for which the piece was originally written and your new audiences and purposes. What changes might you want to make in voice, style, format, and emphasis? Like Sarah's, many of your documents were likely created originally for the classroom. This involved progressive learning and a relationship between teacher and student toward evaluation. Brian D. comments on shifting his view of his audience: "The work itself has obviously matured because beforehand many pieces had the 'this is a college project' feel to them." Although some of these elements are similar, you should try to re-see yourself as a professional moving into an employer/employee or consultant/client relationship. Consider the kinds of changes you will want to make in light of this rhetorical shift. It might mean something as simple as removing the assignment name, header, and professor's name from a research paper or possibly editing to remove grammatical mistakes and spelling errors. It also might mean that the writing would benefit from a change in genre or form. For example, you might recast a paper that was written for a class in an academic format as a feature article or newspaper column. Or you might revise a paper or storyboard into a multimedia piece or PowerPoint format to demonstrate your presentation and design skills.

Revision can take place at many levels. You can look at the maturity of the ideas, the sharpness of the language, the effectiveness of the design, or the quality of the layout. The process starts with reflection and notes. You will need to look closely at your artifacts to determine what you think they need—what is missing or what you wish you had addressed.

When Joy looked through her documents, she realized that she needed to make many changes. This process started with her original resume (Box 4.1), which was outdated and visually bland (See the detailed discussion of revising resumes in Chapter 8.) Joy revised her resume (Box 4.2) to give it a cleaner look, with

BOX 4.1: JOY'S ORIGINAL RESUME

Wanda **Joy** Leake
1000 Crooked Tree Lane
Atlanta, Georgia 30153
707-790-1000 or jleake@verizon.net

Educational and Professional Timeline

2004
Part-time position—electronic editing in MS Word, supervisor, Carol Barns

2001–present
Full-time Student—Southern Polytechnic State University, 1100 South Marietta Parkway, Marietta, Georgia; Major—Technical and Professional Communication (Bachelor of Science)

2000–2001
Student—Forte Fine Arts Studio, Douglasville, Georgia, Owner—Heather Peters; piano lessons

1996–1998
Inventory Manager (printing division)—Ivan Allen Company, 221 Peachtree Center Avenue, Atlanta, Georgia 30101; 404-760-8700; Responsibilities included: Oversee design, production, storage, and distribution of customer forms. Special projects included: Steering Committee for Company-wide Software Conversion and Strategic Planning Committee for Printing Division

1995–1996
Buyer (janitorial supplies)—Atlanta Broom Company, 4750 Bakers Ferry Road, SW, Atlanta, Georgia 30336; 440-696-4600 (now Dade Paper Co.); Special Projects included: Steering Committee for Software Conversion.

1988–1993
Part-time student—Southern College of Technology, 1100 South Marietta Parkway, Marietta, Georgia 30060; Major—Industrial Engineering

1986–1995
Purchasing Manager (store fixtures and supplies), Purchasing Coordinator, Executive Assistant, Word Processor—The Athlete's Foot Group, Inc.—International Headquarters, 1950 Barrett Parkway, Kennesaw, Georgia; Responsibilities included: Oversee design and third party manufacture of custom store supplies, uniforms, and forms. Special Projects included: Steering Committee for Software Conversion

1983–1986
Clerical Assistant (human resources)—Greystone Power Corporation, 4040 Bankhead Highway, Douglasville, Georgia 30134

1981–1983
Claims Manager (3rd party liability insurance)—James P. Daly of the South, Inc. Atlanta Road, Smyrna, Georgia (no longer operating)

1981
Graduate—Lithia Springs Comprehensive High School, 2450 County Line Road, Lithia Springs, Georgia—Honors Graduate; 3.8 GPA—Received the University of Georgia Certificate of Merit

Source: Used with permission from Joy Leake, 2005. Personal information adapted for publication.

BOX 4.2: JOY'S REVISED RESUME

1000 Crooked Tree Lane
Atlanta, Georgia 30153
707-790-1000
jleake@verizon.net

Joy Leake

Traditional Work Ethic, Contemporary Skills

Education
Bachelor of Science, Technical and Professional Communication; Southern Polytechnic State University, Marietta, Georgia; Planned graduation—December 2005—*Magna Cum Laude*

EXPERIENCE

Public Relations—Intern	Debi Curry Public Relations; Marietta, GA; Spring 2005 • Write press releases, Web content, and business correspondence; design graphics (paper and Web); plan and prepare events and the accompanying documentation
Editorial Assistant	Dr. Carol Barns Marietta, GA; Fall 2004 and Spring 2005 • Performed digital editing functions for reference manual conversion to digital format. Part-time position.
Consultant/Analyst	US Office Products, Baltimore, MD; May–Jul 2000 • Strategic planning project; temporary contract position.
Inventory Manager/Buyer	Ivan Allen Company, Atlanta, GA; Jan 1996–Apr 1998 Atlanta Broom Company, Atlanta, GA; Nov 1995–Dec 1996 The Athlete's Foot Group, Inc., Kennesaw, GA; Aug 1986–Nov 1995 • Coordinated the design, production, storage, and distribution of customer forms. Implemented software conversion. • Analyzed, evaluated, and composed fulfillment orders for distribution facility. Improved purchasing process. • Oversaw design and third party manufacture of custom store fixture, supplies, uniforms, and forms. Implemented inventory control system conversion. • Analyzed and maintained optimal inventory levels for distribution center to support over 600 retail stores internationally. Coordinated customer order fulfillment.

TOOLS & SKILLS

Microsoft Office—Expert	Small Group Communication
Adobe Photoshop—Intermediate	Team Collaboration
Adobe Page Maker—Intermediate	Oral Presentation
Macromedia DreamWeaver—Intermediate	Website Development
RoboHelp—Beginner	Online Documentation
HTML—Beginner	Audience Analysis
Adobe Illustrator—Beginner	Document Design

Source: Used with permission from Joy Leake, 2005. Personal information adapted for publication.

sharper attention to document design and layout. She added design elements and color (which is not appropriate in all settings). In addition to changing the design, she worked on the content to remove some of the irrelevant details and address her professional skills more specifically. For example, in the earlier version of her resume she included "piano lessons." Although this is a worthy skill, she chose to shape her resume more precisely and productively toward her field. She also wanted her resume to reflect her years of practical experience in the workplace and the enhanced technological skills she gained by returning to school. She came up with the tagline "Traditional Work Ethic, Contemporary Skills" to demonstrate this philosophy and eventually worked it into her portfolios. She made other small but significant changes such as ensuring consistency in the headings, adding a bulleted list of technological skills, and providing an exit line referring to her portfolio.

Joy made these changes after discussions with and feedback from her colleagues, friends, and an industry mentor. They also resulted from her earlier work on shaping her professional identity.

In another piece, Joy took a rather informal online posting to a discussion board and changed it into a visually attractive article within a totally different context. She extended some of her definitions, added a table for readability, and revised the language to address an audience that was not part of her existing online community. She used many of her original ideas but recast them as a critical review or magazine article complete with a heading, byline, and pull quotes. At first glance, she might have cast this piece aside, thinking it was inappropriate for inclusion in her portfolio. Instead, she realized that she liked what she had said but was uncomfortable with the format and presentation. Recasting it was the answer.

In a similar genre shift, Joy returned to a piece of writing that she produced before her degree program in a survey literature class. She decided to recast it as a literary review in which she compared two Southern authors, Eudora Welty and William Faulkner, creating a stylish review that mimicked the design conventions of other literary reviews (Box 4.3). She created two columns and used a watermark with the state of

BOX 4.3: SOUTHERN LIVING

Both William Faulkner and Eudora Welty were from Mississippi. They used this region of the country (the deep south) as a backdrop for their stories; Faulkner a fictitious town of Jackson based on Oxford, MS and Welty used the undisguised Natchez Trace. Welty used place and setting more than action, to tell her story and Faulkner used social conditions and tradition to tell his.

An early interest in photography influenced Welty's literature by enabling her to linguistically paint the scene for the reader. "It was a shotgun house, two rooms and an open passage between, perched on the hill. The whole cabin slanted a little under the heavy heaped-up vine that covered the roof, light and green, as though forgotten from summer. A woman stood in the passage." This scene is a typical mountain dwelling from somewhere in north Georgia or Tennessee. From the description of the simple dwelling, the reader is pre-disposed to the simple people living there.

Faulkner used the social traditions of the south to tell his stories. The pressures of knowing what was expected of Emily, in *A Rose for Emily,* drove her to take certain action. For example, a woman was expected to be married before 30 years of age; "When she first begun to be seen with Homer Barron, we had said, "She will marry him." Then we said, "She will persuade him yet", because Homer himself had remarked—he liked men... Later we said, "Poor Emily". Finally the women of the town thought it disgraceful for them to date and sent the minister to intervene first and then her relatives. The pressures they put on her left her no choice, in her mind, but to fake a marriage to Homer and murder him to keep him with her.

While both writers used the south as the setting for their stories, they each used different methods to bring the reader into that setting. Faulkner used psychological conflict; when the heart conflicts with traditional expectations and Welty used extensive description; a verbal painting.

Mississippi (home state for both authors) in the background, added photos of both authors, and included pull quotes from their works. She also included a list of references that helped the article stand on its own outside of the classroom context of the literary response journal. She can now use this revised document (Figure 4.1) that, again, might have been discarded for irrelevance, to achieve her purposes as she demonstrates sharpened skills at the same time.

Angela, a professional working for a broadcast company, took a document that was created on the job and applied her new skills in her revision. This technical procedure document, written mostly in narrative format, was restructured to make it more user friendly. In the new document, titled "How to Correctly Time NBA HD/SD Audio," Angela reorganized the procedure into chronological steps and included graphics for each step to provide visual instructions. Through this revision she demonstrated not only her rhetorical awareness of the audience, but also her on-the-job experience and her knowledge of usability and procedural documents. In addition, she showcased her use of graphics through screenshot photos she created.

FIGURE 4.1
Joy's Revised Literary Analysis
Note: See Box 4.3: Joy's Original Literary Analysis.
Source: Used with permission from Joy Leake.

Another student, Wylie, explains the processes he used in revising a text-only assignment for an introductory course in his technical communication major. The assignment was originally written as a rather informal response to prompts about industry standards. Wylie speaks about his writing and design processes:

> Before I could add images and formatting changes, I had to rewrite certain parts of the assignment that only the professor would understand. Once I changed the aforementioned parts to be more easily understood by any audience, I had to make the documents more appealing to the eye. I added pictures of the phone and menu system for the "Bookmark Blues" article. This helped the piece look more like a magazine article than a school assignment.

As demonstrated in Figure 4.2, Wylie concluded that the most effective revision of this document addressed issues of format, audience, and graphics. He also took the opportunity to learn new a software application in order to include attractive and advanced features such as borders, screen captures, and digital imagery.

REVISING FOR CONTENT

You cannot separate context from content. It is important to think carefully about the appropriateness and depth of your ideas. Readers will be drawn to interesting visuals, but the portfolio also provides a place for you to assess your knowledge, philosophy, writing skills, and work ethic. The images you portray through the subjects and content included in your portfolio require careful consideration. You want to be seen as professional and knowledgeable.

One level of content concerns the affective domain. Consider how others will perceive you through your ideas. For example, you might think about the differences between disclosing in a classroom and in a professional community. Unfair as it might seem, people are always affected by biases and ethics. Make sure that you do not offend others or disclose more than you should in a professional setting. On the other hand, you might consider it important that employers immediately understand personality traits like Trina's willingness

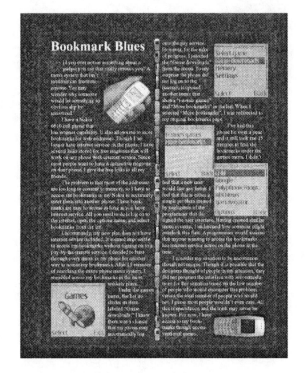

FIGURE 4.2
Wylie's Revised "Bookmark Blues" Article
Source: Used with permission from Wylie Jones.

to speak out. Another student, Angela, initially resisted self-disclosure in her portfolio. This became a content issue for her as she questioned the inclusion of a piece that alluded to her more radical side as she designed a creative resume equating herself with a famous activist of the 1960s. She felt that some audience members would not understand the context or the humor. Angela explains:

> I noticed that I had to discard my reservation of excess personal disclosure when revising and choosing my pieces. Specifically, I decided to use my creative resume after months of absolute resolve against it. I felt that certain audience members would feel challenged by the humor.

Angela ultimately decided to include the piece in her portfolio. Readers' reactions to it confirmed that she had made the right decision. As she says, "To my utter surprise, that particular document received a lot of praise accompanied by chuckles." She chose to take a risk in showing off her creative skills. However, in another document, Angela came to a different conclusion. Here she was concerned about the critical nature of her ideas. Although she originally prepared a PowerPoint presentation to emphasize racial inequity in a small Southern town, she realized that some audience members might not share her philosophy or might believe that her opinions were too extreme. She knew she wanted to use this document because it demonstrated her presentation and research skills. She eventually chose to edit out particularly graphic or extreme images and tried to reframe it using a more neutral voice. Maintaining the integrity of the piece, Angela kept the figures and statistics—which added to her credibility—and toned down the emotional appeals that she thought might offend others.

Another student, Judith, reconsidered the ways she had used personal examples in essays written for class assignments. When revising a document on gender differences in communication, she considered changing the ways some of her arguments were framed. She explains:

> In my document, "Battle of the Sexes," I had used many personal examples within the original text. I couldn't decide whether I should keep them in and if I did I wondered how that would affect my audience. I ended up keeping some of the personal examples because they strengthened my argument, however, taking them out when I felt like they didn't. I left in personal references to my husband because I felt the article worked better with my use of personal analogy.

Even though it revealed something about her personal life—that she was married, as well as the nature of discussions in her marriage—she chose to include the information because it strengthened her claims.

In addition to issues of disclosure, the portfolio can demonstrate your knowledge of your field and industry standards. Consider how you come across as a writer in terms of content, style, voice, organization, and research. Establish your credibility through content as you include current issues in the field and demonstrate knowledge of field conventions and competencies. Carefully cite outside sources to demonstrate ethical research skills. Consider every piece a showpiece for these skills and characteristics. Your portfolio pieces should ultimately reflect your depth of thought and knowledge in the field, along with the effectiveness of your expression.

AMPLIFYING: REPRESENTING MULTIPLE SKILLS WITHIN A SINGLE DOCUMENT

In the diverse field of communication, employers are looking for a demonstration of multiple skills. As you revise your documents, consider how multiple skills can be represented in a single document or collection of work. Whereas some of your early drafts used word processing and writing skills, you can now also incorporate what you know about design, layout, through document design. This relates back to the concept that portfolios are representative rather than comprehensive. Sheer volume is not enough. In fact, as we have discussed, in an interview, you might be able to show only a couple of documents. Both depth and breadth can be demonstrated when a single document provides many talking points regarding your skills. Think about how one document might showcase your writing, design, and even interpersonal skills. As you look through your work, reflect on the ways that your pieces might demonstrate more skills than they did when originally created.

Remember, your portfolio is a visual document that you hope will initiate discussion. Images can draw your audience in. A research paper written for a class on the "shock strategy of advertising" can be greatly enhanced by including a few pictures of actual campaigns. You might consider the ways a personal piece, such as a literacy autobiography or memoir, might achieve a different effect if images such as book covers, scanned illustrations, childhood photos, or original digital pictures are included. You can enhance or amplify your drafts through illustrations, charts, and figures as well as with layout, design, and font choices. Genre imitation—such as the two-column, pull quote convention used in newspaper and magazine articles—can provide productive revisions and make your work appear professional. Your revisions might also address your increased knowledge of a particular software application or communication skills such as organization, teamwork, and sharp writing.

Wilda made major changes to a paper that she originally wrote for a Health Management class, deciding on several things that she wanted to communicate. She describes her revision decisions:

> I wanted to incorporate some of my management classes into the portfolio, and being that I worked for the CDC [Centers for Disease Control and Prevention], I thought this was the perfect paper. This, on the other hand, turned out to be my biggest challenge. I added graphics and color to make it look like a brochure, but that didn't work. In the next revision, I changed it to a pamphlet, had to change some of the wording because some of the information was offensive to my reviewing audience. In my final draft, I changed the colors and added blocks and graphics that illustrated the types of people. I kept the bullets consistent, and it came out to be the best professional pamphlet I had ever created.

Figure 4.3 shows how Wilda transformed her original research paper into a pamphlet. She considered the issues of length and readability for an audience from the CDC. It was not likely that a CDC client would pick up and read her original ten-page research paper. She realized that a different format could connect more effectively with her audience—a shorter, three-fold brochure that clients could easily scan in a waiting room. She changed the wording and condensed the information from a narrative format into a bulleted list. In addition, she included pictures to make it an attractive document that people with mental health issues (sources of potential social discomfort) would feel comfortable reading in a public setting. At the same time, she demonstrated her knowledge of technology, research, document design, and graphics, along with field-specific knowledge.

EDITING

The last issue to consider in your local revision is editing. As the definition provided at the beginning of this chapter states, editing concentrates on surface features and overall consistency. These might include

* Spelling
* Punctuation
* Word usage
* Grammar and syntax
* Consistency and style
* Format

An excellent local editing strategy is to read your writing aloud and correct as you go. It is amazing how much self-correction can take place when you actually hear your words and notice how they work in speech. You might also ask a mentor or friend to scan your documents for mistakes. Sometimes we get so involved with our own work that it is difficult to catch small or pattern errors. Obviously, you should use the technological devices available for editing, such as spell checkers (critically examine your computer's suggestions,

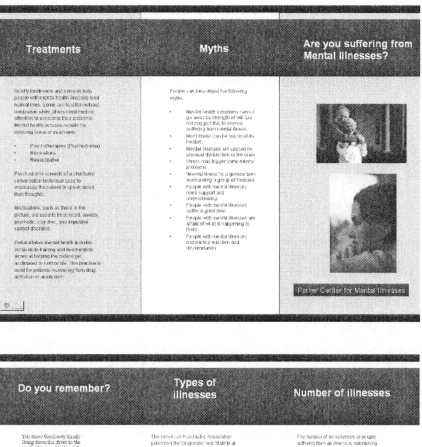

FIGURE 4.3
Wilda's Revised Mental Health Pamphlet
Source: Used with permission from Wilda Parker.

but make your own final decision as a writer). As Judith comments upon reflecting back on work produced earlier in the program:

> Reviewing my work and changing it for portfolio quality was eye-opening. I had to check for grammatical errors, of which I found more than I expect. More importantly, I checked for content errors, and I had to change something about pretty much all of my documents.

EXERCISE 4.1 REVISING FOR PORTFOLIO QUALITY: REVISING LOCALLY

After pulling possible pieces together for your portfolio, you will be faced with the task of selecting and revising particular documents (local revision). Portfolio-quality revision involves seeing your assignments/projects in a whole new way. It is necessary to reconsider your rhetorical situation as you shift your intentions, purposes, and audiences in your professional portfolio. You will also need to carefully edit your texts to eliminate grammatical, mechanical, and spelling errors. Your ultimate goal here is to create a finished, professional document that represents your skills as a writer and a designer.

Choose one piece of writing that you definitely want to include. Use this revision assignment as a model for revising the rest of the documents in your portfolio, taking a written document and revising it for portfolio quality (including the incorporation of images). Here are some issues to consider:

- **Audience:** You are moving from a classroom/teacher audience to a professional/workplace audience. Consider the changes that might be needed in voice, style, content, language, length, and so on to communicate effectively with this audience.
- **Purpose/Skills:** Think about what skills and purposes you want to convey. Consider the ways you might revise the document to communicate them more effectively.
- **Subject:** Although your subject and multiple perspectives remain the same, think about how your format will change. In the original paper, you probably worked to elaborate, complicate, and extend your thinking. In this version, you might want to boil your ideas down—for example, presenting complicated data in a chart or a bulleted list. This will preserve the rich meaning of your paper and at the same time make it a more readable document. You should still completely cover your subject, as well as (of course) cite and document your sources.
- **Format/Style/Document Design:** Many of your assignments are in an academic format such as a paper or other typed work. Consider altering the style/format to that of a newspaper or magazine article. Work with white space, fonts, pull quotes, headlines, and so on to demonstrate your ability as a writer and a document designer. Your skills with different software applications will also be highlighted.
- **Images:** The portfolio is a visual text. Pull images into your writing assignments in the form of photos, clip art, or illustrations. Draw upon the images that you found and took for a freewrite assignment. Caption your images carefully so that they enhance your reading.
- **Copyright:** How do copyright issues affect the choices in your portfolio? Check the source of all your images, tables, and graphics. Make sure that you are not using copyrighted material without permission (see Chapter 6 for a detailed discussion of copyright issues).
- **Context:** Many of your writing assignments and documents do not stand on their own outside of the courses in which you completed them. As a writer, it is your job to find ways to contextualize your writings so that they can be viewed in isolation (and so that the reader understands the skills that went into creating the document). This might mean including abstracts, changing introductions, or describing the context in a short statement.

Note: You should conduct a short review of magazines and articles to get a good sense of the genre and possible choices for page setup, borders, color, picture and pull quote format, fonts, and so on.

REVISING GLOBALLY

Once you have revised your texts, it is time to start thinking about global revisions and the shape of the whole portfolio. As defined at the beginning of this chapter, global issues "deal with macroscopic issues in the text or global matters" such as ideas, organization, or overall theme. This is where you work in your theme, design, visuals, and context across the portfolio and explore the connections between your artifacts to produce a coherent, whole body of work.

Organization

Global revision involves looking at the connections between ideas, sections, and documents. The organization of your portfolio takes into account its structure as a whole—how artifacts are clustered and how the sections relate to one another. Sections can be based on content, theme, genre, or chronological progression. Readers often cannot process large amounts of information at once. Imagine trying to read a novel without chapter breaks! Information is easier to understand if it is broken down into smaller units. Also, the portfolio can be used to tailor your skills to different kinds of job descriptions. It is important to be able to move quickly from one section to the next or eliminate sections that do not apply to particular jobs.

In some cases, your artifacts will be grouped because of their similarity, such as a series of brochures that you wrote and designed for a single marketing campaign. For example, Joy created several pieces for a nonprofit organization called PAWS (Progressive Animal Welfare Society: http://paws.org). At one point, she thought about spreading these pieces throughout her portfolio under the particular genre (Web site, brochure, presentation). Instead, she decided to cluster them as a "Marketing Campaign" (Figure 4.4) to demonstrate flexibility, consistency, multiple design, and technological skills along with the interpersonal skills of public relations and teamwork.

Joy also grouped her work according to skill sets and types of documents. She describes her strategy for organization:

> I grouped the examples into business sectors. For example, I used Marketing and Promotion (graphic design and brochures), Document Design (technical documents), Web sites (web design). I imagined that, based on the job I was applying for, that particular section would be focused on in the interview.

In other cases, you might want to showcase particular skills together or focus on a specific subject such as medical writing or international communication. Michel, for example, included a newsletter, an article, and a Web site on one disease, tularemia. This highlighted the skills needed in medical writing that requires communicating with different demographic groups and audiences. Your portfolio design and table of contents completed in Chapter 3 should provide guidelines as you start the process of global revision.

Although you are extremely familiar with your work, your audience is not; in fact, they probably are looking at it for the first time. This raises issues for you as a writer. What do you have to do to immediately

FIGURE 4.4
Joy's PAWS Marketing
Campaign
Source: Used with permission from
Joy Leake.

capture their attention? How do you wish to guide their reading of your portfolio? What context does your audience need to understand your intentions and assess your skills?

Integrating theme and consistency

In your portfolio design proposal, you should have come up with a unifying theme. In what ways will your theme, metaphor, and design elements work throughout the portfolio? Where do you see consistent connections between color, ideas, and design elements? As we discussed earlier, for some of you this might mean a dedicated symbol, philosophy, or personal interest. For others it might be a more subtle connection that involves design elements or graphic markers. Either way, you will now revise to increase this consistency and communicate your objectives. There are many issues to consider, such as how to implement your theme and what types of paper or electronic resources are available.

It is important to carry your theme or metaphor throughout your work, providing a thread that holds it together. Wilda decided on the theme of books because she "always loved books" and wanted to "be a librarian one day." Although she was not immediately directing her portfolio to library science, the book image provided a visually appealing way to work subliminally toward that goal and emphasize her school-work and knowledge. She used the book as an organizing principle for the global structure of her portfolio, combining a picture of an old-fashioned book and the title "Library of Knowledge." Wilda explains this connection:

> The gold in the book was the primary color, and from there I used any color that had a gold tone to it. My sections were divided up like a chapter in a book, and each chapter had something different to offer.

Angela chose a theme of light, with each section beginning with some depiction of light along with an appropriate quotation. She explains the connections between her theme and the overall layout of her portfolios:

> I was able to use my need for honesty as an asset when communicating my theme of technical communicators as "bringers of light or clarity." Each section of the portfolio begins with some depiction of light while emphasizing the correlation between light and truth through an accompanying quotation.

She recognizes the transformation of light as an arbitrary symbol into an embodiment of her philosophy and skills:

> My original light metaphor was merely a recurring visual used to provide continuity. As I continued to revise my individual documents and globally revise the portfolio, my theme transitioned into a more accurate representation of my professional identity.

Wylie took a different direction, using the magazine format as his common thread. He wanted to demonstrate that "uniqueness is [his] most valuable asset" and created an "off-the-wall idea of a magazine for his paper portfolio." Rather than placing his documents in a binder, he imitated the style of *Newsweek* magazine. He says:

> I wanted to create something that would not be so imposing and would set the readers at ease and not put them on edge. Not that many people have seen the big leather portfolios. Everyone has seen a magazine. People fear the unknown. I want my interviewer to see me first and then the fact that I am carrying a portfolio, not the other way around.

His global revision involved studying and carefully recasting his work in a magazine format, including letters to the editor, mock advertisements, book reviews, and feature stories. He even included an advertisement for a movie on the back cover to refer to his filmmaking and multimedia skills. Wylie speaks about the processes he used and the way he constructed a model to help him organize the work:

> *Newsweek* had roughly a one-to-one ratio of ads to stories. I tried to mimic that idea by including an ad on one page and a story on the opposing side. The feature article, though it didn't have an ad, had a separate story spread across the bottom of both pages, much like the feature article in the magazine.

In addition to the model, Wylie worked to tie in underlying messages. For example, his front cover featured an image of himself (Figure 4.5), with many arms holding different symbolic artifacts (computer, pen, cell phone, camera, etc.) representing his flexibility and knowledge across the field. Here he describes this image:

> This project gave me a greater respect for many different jobs. I acted as the photographer, editor, design team, draftsman, column writer, feature writer, copy editor, general editor, page layout specialist, printer, publisher, and more. The many arms on the cover represent, in part, the notion that I had to play all of these roles to complete this assignment. I feel this indicates that I could go into any of the aforementioned careers.

Because Wylie chose a format as his guiding thread, he also found himself generating new material to imitate the style and content of the magazine. He explains: "I originally wanted to use more content from the past, but I just didn't like the way some of my old documents would fit into a magazine. I started creating documents from scratch."

Sometimes students use a dominant symbol as a connecting thread. Joy, for example, used the image of a kaleidoscope as a theme for her portfolios, focusing on the triangular shape that created the images and bright colors. She explains:

> I have a multifaceted approach to problem solving and wanted to illustrate that with the examples in my portfolio. But since I am also creative, I wanted to use bright colors. I used six separate colors, one for each section, faceted backgrounds, and the triangle shape throughout to keep the theme consistent.

Each colored section corresponded to a document division. This theme and structure were used in both her paper and electronic portfolios. Similarly Judith, also used color as a dividing strategy when organizing her materials. The colors became her theme, as each one held particular significance for her. She says:

> My metaphor is really the colors that I chose to represent me and my work. I chose red to suggest passion and vitality in which I placed my articles and reviews, blue to express more professional documents which I used for my document design section. Orange represented warmth and creativity, in which I placed my graphics. My colors perfectly matched the items that I placed within them as well as aspects of my personality.

MORE THAN A CONTAINER: CONTEXTUALIZING ARTIFACTS

The last level of revision concerns context elements that "go beyond the immediate concerns of the text and deal with the contexts that inform the piece of writing" such as rhetorical context, physical or social setting, or industry standards (Straub and Lunsford 1995, 159). As should be obvious by now, your portfolio is more

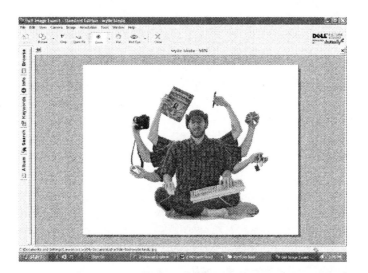

FIGURE 4.5
Wylie's Multi-Armed Image
Source: Used with permission from Wylie Jones.

than a container. It should tell a story as it demonstrates your identity, skills, and progress. Here is where the theme or metaphor helps you to shape your story and contextualize your work. Imagine the concerns of your audiences when reading your work. What outside factors influence that reading, and what might they need to connect your artifacts together to form a comprehensive body of work? It is up to you to contextualize the portfolio for the reader. We suggest you achieve this through context statements that guide the reader's attention toward your thematic connections, processes, and skills.

OVERALL STRUCTURE AND CONTEXT

Many of our students found it helpful to think of the portfolio as a book or an inclusive document with a table of contents, an introduction, a body, and a conclusion.

Table of contents

In your paper portfolio, the contents should take the form of a table of contents; in the electronic version, it will most likely be your home page. It should outline the organizing structure of the portfolio and guide the reader by indicating section divisions, types of documents, and titles of particular works.

Introduction

The introduction tells the story of your portfolio. It might consist of a personal statement that welcomes readers and guides them through the portfolio. It might also include information about your background, the processes you used in creating the portfolio, an explanation of your theme or metaphor, and instructions for reading or navigation. You can also include information about particular software applications that might be needed to download and read the electronic version.

Your theme or metaphor depends on the context to enable readers to read the portfolio the way you intended. A rock, for example, can evoke an image of strength for some readers or something primitive or dangerous for others. As mentioned in Chapter 3, it is your responsibility to make the theme your own by naming it and connecting it to your purposes. Introductory context statements can help get your ideas across. Brian D., whose metaphor was the marathon runner and the track, explains the way he presents these images in his introduction:

> Technical writing is like a marathon with many important milestones to achieve in any given project. And it's always a race against the clock. When the starting gun goes off, a technical writer needs to be ready to get the project going and see it through to the end. This is no simple task. Training, skill, and talent will all be tested. A good technical writer will get the job done right and get it done on time. A good technical writer will go the extra mile.

In this introduction Brian sets the pace by indicating that he understands the demands of the technical writer. He carries the theme throughout but keeps it focused on his purposes. Similarly, Amy describes her puzzle imagery in her introduction:

> There are so many different things that make up who I am. You could ask people who know me what things make up Amy, and they would tell you that I am like a puzzle. There are some people who never get to see the whole picture. Some pieces I choose to hide. Some of the pieces that people see in a social setting simply do not come out in an academic setting.
>
> This portfolio helps to bring these pieces together. I have samples from the things that interest me and that help to define who I am. You will find work I have completed for WGHR, my campus radio station. You will also find work I created for Gamma Phi Beta Sorority, of which I am a member. There are examples of things I have done for my job as a Resident Leader and Tutor. There are also pieces from subjects that interest me, such as the

medical field and my hobby of soap making. The subject matter covered in this portfolio reflects the variety of things that interest me.

Joy, who used the metaphor of the kaleidoscope, needed to explain the complicated way she connected it to her goals and objectives. She includes a diagram of a kaleidoscope and explains the way the "three mirror kaleidoscope makes an endless field of patterns," with the three sides labeled "experience," "creativity," and "education." She elaborates:

> Like a three-mirrored kaleidoscope, I reflect on information from several perspectives utilizing education, experience, and creativity. This enables me to find solutions, organize information, and communicate more effectively.

Like Brian D., she uses this context statement to connect her theme with her skills and purposes. To emphasize this even further, Joy designed a logo (Figure 4.6) for her portfolio that included the kaleidoscope image along with her tagline, "Traditional Work Ethic, Contemporary Skills." She repeated this logo throughout the portfolios, reinforcing her message and branding her name. She also included the logo on a CD cover and a business card to leave with potential employers.

Sometimes readers need more explanation of the particular fields or terms. For example, Sarah defined "international technical communication" in her introduction as follows: "International Technical Communication is applied communication to allow clearer understanding of all aspects of technology through graphics, design, sound, video, and writing in an international environment." She considered this necessary because it emphasized the uniqueness of her degree, which might not be well known to others.

Miranda provides a final example as she connects her portfolio to her Art Deco theme, describing her skills, design decisions, and connections. Here is the introductory context statement included in both her paper and electronic portfolios:

> Welcome to my portfolio of user-centered information design. I built this portfolio to showcase my skills and experience in the field of Technical and Professional Communication. I chose the sleek style of Art Deco as my theme because the culture that inspired it reminds me of the principles taught in this field.
>
> In the Industrial Age, speed was the metaphor for modern times. From this came the Art Deco movement—a movement towards simplicity, streamlining, aerodynamics, clean lines, and vivid colors.
>
> Now, in the Information Age, it is the speed of information that is most important. In my field, we use principles in usability, readability, and design to "speed up" communication. Importance is again placed on simplicity, streamlining and clean line in this fast-growing profession.

FIGURE 4.6
Joy's logo
Source: Used with permission from Joy Leake.

Miranda also does a good job of directing her readers in her electronic portfolio, including the following instructions along with her narrative introduction:

* Please choose one of the three areas from the tabs above.
* For best viewing set your screen resolution to at least 1024 × 768 pixels.
* Set your color quality to at least 16 bits.
* Have a PDF reader installed capable of reading Adobe PDF files 4.0 or higher.

These instructions help the user to view the portfolio the way Miranda intended. There is nothing worse than a potential employer opening up an electronic portfolio and being unable to read the files or reading them incorrectly. These instructions ensure consistency on different computers and systems.

A final note on Miranda's opening page is designed to protect herself and her material (See Chapter 6 for a full discussion of legal and ethical issues affecting portfolios.) She includes the following copyright statement:

> Copyright 2004 Miranda Bennett. All rights reserved unless otherwise indicated. All materials on these pages are copyrighted by Miranda 2004.

Keep in mind, however, that the portfolio is an evolving document and that you will need to change the copyright statement each time you add items.

As we all know, first impressions count. Your introduction or home page will be the readers' first glimpse of your portfolios. It sets the stage for the rest of their reading experience, introducing your look, style, and perspectives, so work to carefully craft something that is clear, attractive, and inviting.

Section introductions

Like the whole portfolio, each section needs an introduction. These introductions provide transitions to other sections and maintain thematic consistency. When dividing your work into sections and categories, it is important to guide readers along (in both paper and electronic formats). This means thinking carefully about technical issues such as navigation and captioning, as well as design elements that visually divide one section from the next. It also requires a narrative element that explains the section's purpose, skills, and connection to your theme. Most of our students also included a visual cue (color or graphic change) that indicated movement from one section to another. In these section context statements, you might also add a section table of contents or highlight the skills used in the particular section.

Brian D. continues his running metaphor, with each section represented by a mile marker that includes different kinds of writing. Here is an example of one of his section context statements in which he explains instructional writing:

> It's still early in the race, just far enough in to start settling into an easy, relaxed stride. Here's where all the hours of training start to pay off. All the instructions and advice given by coaches and other experienced athletes really start to make sense. The coaches knew what you'd be thinking and how you'd be feeling at this point in the race, and you're benefiting from it now. You're six miles into the marathon and distracted by a million different things, so the instructions had to be clear and simple. That's how it works with really good instructions. A good technical writer can anticipate what the reader will be thinking at each step of the procedure, and knows what information the reader needs to get to the next step. A good technical writer also knows what information the reader doesn't need, information that will just be one more unneeded distraction for a busy reader.

He clearly demonstrates his knowledge of instructional writing in this statement and follows it up with a brief annotated list of the section contents right on the page. Using this list, readers of the electronic version can link directly to the documents.

Joy, on the other hand, created section context statements that were mostly informational. For example, her section on Marketing and Promotion says:

> For a marketing program to be effective, the theme and message must be consistent and unified. These are examples of campaigns that I have created or helped to create.

She also included a pertinent quotation in each section, such as this one from Mark Twain's book *A Connecti-cut Yankee in King Arthur's Court*: "Many a small thing has been made large by the right kind of advertising."

Other students, like Angela, drew upon compelling quotations referring to light (her theme) to open up each of her sections:

Document Design: In the right light, at the right time, everything is extraordinary.   Aaron Rose

Writing: The difference between the right word and the almost right word is the difference between light-ening and lightening bug.   Mark Twain

Graphics: Work while you have the light. You are responsible for the talent that has been entrusted to you. Henri Frederic Ameil

Accolades: There are two kinds of light—the glow that illuminates and the glare that obscures.   James Thurber

Angela also used these section dividers to showcase her photography skills and included her own pictures along with a short caption on each one, following every caption with a short explanation of the section and the tools used to create the documents. Angela says, "This document design section demonstrates the effec-tiveness of clarity, graphics, layout and text."

It is important to communicate your skills throughout the portfolio. Although many of your readers will know immediately what tools you used, we find that it is best to name them in these context statements, either as part of your narrative or simply in list form.

Document context statements

Your readers must not only understand the connections and themes throughout your work, they must also be able to glance at the individual artifacts and get a sense of the skills and purposes behind them. Once again, readers or users may skip around in the portfolio, so each page should stand on its own and connect to the larger body of work. This means making rhetorical decisions about issues such as labeling, captioning, and individual listing of skill sets. A context statement should be created for each document. It can be as simple as a label or link, or a short explanation of the skills or software used to create the document. In their electronic portfolios, many of our students also used thumbnail versions of their full documents along with explana-tions. These guideposts should appear throughout the portfolio so that readers can turn to any page, view it in isolation, and at the same time connect it to the larger body of work. For each of her documents, Judith included a title and a short explanation demonstrating her skills. Here are two document context statements from her graphics section:

Robot Waitress: The Robot Waitress was created using Adobe Illustrator. The complexity of this graphic helps to give it a cartoon-like character.

KSOL Rising Radio Station: This ad was created in Photoshop. I utilized a variety of different tools within the software to create the blending effects of the radio emerging from the water.

Figure 4.7 from Judith's electronic portfolio shows how she designed these context statements to include a thumbnail of each image along with a detailed explanation.

As you can see from the preceding examples, document context statements reveal many different things to readers. You can use them to discuss the tools you used, the skills involved, or even the roles you played in team projects.

Sarah uses her document context statement on South Korean business practices to show her involvement in a group presentation:

South Korean Business Practices: A group presentation, given like a seminar, speaking on the various aspects of conducting business in South Korea. My section focused on the religion, social customs, and psychological methodologies of the South Korean business person. I was also the moderator for the group. The main tool used was PowerPoint.

FIGURE 4.7
Judith's Document Context
Statements
Source: Used with permission from
Judith Dickerson.

She finds an ethical way to describe her contributions to this project. These statements might also include other credits or references to published work, such as Joy's context statement for her travel column about camping in Florida:

> This travel article that I wrote in March of 2005 was published in the *Huntington Connection Magazine*. It has a circulation of approximately 50,000 in the Long Island, New York area.

The statements might also reveal particular likes or dislikes of the writer, as in Joy's description of her literary review "Savoring the South." Her document context statement reads:

> I have always enjoyed reading works of Southern authors. This article compares the styles of Eudora Welty and William Faulkner, both from Mississippi. I used Photoshop (graphics) and PageMaker (layout) to put this one together.

Through this statement Joy alludes to her interest in Southern literature and the South in general, giving the reader a brief summary of the longer text. Like other students, she chose to include the tools used to create this document.

All of these statements represent a few ways of defining and speaking about your work. Like the whole process of creating the portfolio, they give you a chance to articulate your learning and skills to others. Through the introductions, section statements, and document context statements, readers should be able to navigate your work with ease.

SUMMARY

Although revising your work to achieve portfolio quality is not an easy task, we feel you will find it rewarding. Our students speak of the amazing transformations that occurred as they applied their new knowledge to their old works and revised them for new audiences and purposes. Chapter 4 provides strategies for revision and the creation of context statements for your portfolio as a whole, including introductions, section statements, and document context statements. Through both local and global revision, you can create professional documents and a comprehensive body of work that reveals your skills, goals, and identity. Judith sums it up nicely: "By performing simple changes to my work and altering them toward portfolio quality—by adding graphics and presenting them in a way that draws in the reader—I made a tremendous difference."

ASSIGNMENTS

Assignment 1: *Revising for Portfolio Quality—Revising Globally*

Global revision involves considering the way your portfolio works as a whole, including organization, theme, and context. Complete these assignments for both your electronic and paper portfolios.

PART 1: ORGANIZATION AND THEME

This is a 10-minute writing assignment in which you examine the following questions:

* How do you plan to use design elements (graphics and color) to create consistency throughout your electronic and paper portfolios?
* How will you carry your theme throughout both portfolios?
* How will you divide your portfolio into sections?
* How will you group your documents within each section?

PART 2: CREATING CONTEXT

Draft context statements for the following elements:

Introductory context statement

This statement should tell the story of your portfolio and include items such as an overall readers' guide and an explanation of thematic elements, technological cues, and opening credits.

Section context statements

These statements should describe your section divisions and summarize the appropriate section. They might include elements such as section contents, visual elements, outside resources, reader guidelines, and thumbnail links (for the electronic version).

Document context statements

These statements should describe particular documents, including their purposes and background, along with the tools and skills used to create them.

REFERENCES

Bishop, Wendy. *Acts of Revision: A Guide for Writers*. (Portsmouth, NH: Boynton Cook), 2004.

Straub, Richard and Ronald F. Lunsford. *12 Readers Reading: Responding to College Student Writing*. Creskill, NJ: Hampton Press, 1995.

Twain, Mark. A *Connecticut Yankee in King Arthur's Court*. New York: Signet Classics, 2004.

The Electronic Portfolio

E-portfolios are so versatile. As you create artifacts, you can add them to your portfolio at virtually no expense, except your time. I carry electronic copies of my portfolio on my jump drive and link new documents to my site. While I was creating my portfolio, I was teased for wearing my jump drive on my neck as if it were a locket with a photo of my mother. Joy

INTRODUCTION

This chapter introduces the electronic or digital portfolio (e-portfolio) and describes its purposes, strengths, and contents. You will be guided through the process of planning and putting together an e-portfolio. Chapter 5 covers the following topics:

* Defining e-portfolios
* Purposes and advantages of e-portfolios
* Contents of an e-portfolio
* Planning e-portfolios
* Designing e-portfolios
* Creating templates
* Adding graphics
* Using Portable Document Format files
* Saving your e-portfolio in HTML format
* Adding auto-run to your e-portfolio
* Adding audio to your e-portfolio
* Uploading your e-portfolio
* Getting search engines to find your e-portfolio

You will also learn how to save your e-portfolio in digital format. In addition, you will find information on how to upload your completed e-portfolio to the Web and learn how a search engine finds this portfolio.

Pages from an SPSU Student's E-Portfolio

DEFINING E-PORTFOLIOS

You may have already guessed that e-portfolios are nothing more than portfolios in electronic format—that is, digital portfolios. The main difference is that they are designed to be viewed on a monitor rather than on paper. E-portfolios usually contain the same artifacts as print portfolios, but they may also contain hyperlinks, interactive graphics, sound, and video. So, while the main contents of the paper portfolio and the e-portfolio are basically the same, the way the e-portfolio is presented sets it apart. In addition, while the audience for the paper portfolio is usually limited to you or a specific group chosen by you, the audience for the e-portfolio is usually intended to be much larger. Your e-portfolio can virtually publish your work to the world, whether the quality is excellent, good, or substandard. This is the major reason to give careful consideration to the creation of your e-portfolio. You must be sure that the portfolio entries represent only your best work.

PURPOSES AND ADVANTAGES OF E-PORTFOLIOS

Purposes

Your e-portfolio will have many of the same purposes as your paper portfolio, but it will also offer several advantages. Chapter 1 stated that portfolios are classified as either working, academic, assessment, or professional. According to Kimball in *The Web Portfolio Guide: Creating Electronic Portfolios for the Web*, portfolios may be either personal and informal, which he calls "working portfolios," or selective, revised, and formal, or "presentation portfolios" (Kimball 2003, 8). E-portfolios are both professional and presentation portfolios. Your e-portfolio will present to the professional world the work that you believe qualifies you to be a part of that world. Paul Treuer, Director of the University of Minnesota–Duluth's Knowledge Management Center, says, "The whole idea behind these e-portfolios is to give others a complete sense of what you're all about" (Villano 2005, 45).

The e-portfolio provides portability that the paper portfolio lacks. The paper portfolio may be one-of-a-kind, holding original drawings, paintings, diagrams, and documents, or perhaps duplicated entries to produce multiple copies of the portfolio, but the end result can be restrictive; the portfolio can be large, expensive, hard to carry, or difficult to mail. The e-portfolio, on the other hand, can be duplicated, shipped, and electronically shared inexpensively with anyone who can view computer files. You can e-mail it at no charge, or you can copy it to a CD or DVD and mail it for slightly more than it costs to mail a letter. It can hold the complete contents of a long book, a large collection of drawings or photographs, samples of film clips, or a combination of all of these artifacts. Whereas your paper portfolio's linear page-by-page structure restricts the flow, hyperlinks in the e-portfolio make it possible to organize your entries in several different ways at once. Your audience can then choose, based on individual interests, several different ways of looking at your e-portfolio.

Advantages

Finally, designing and distributing your e-portfolio might be the perfect way to distinguish your resume from the piles of paper resumes crossing a potential employer's desk. A good e-portfolio not only publishes your work for others to see, but also shows off your electronic and organizational skills. "Sometimes, it could pay to be a little different," says Maria Mallory White, writer for *The Atlanta Journal-Constitution*, in her article "Break Out of the Standard Resume Mold." She quotes Georgia Tech Director of Career Services Ralph Mobley:

> We are challenged to "think out[side] of the box" so often it has become a tired cliché. . . . Yet, in at least one key area of our work life, the job search process, we have eliminated the very individuality and creativity [our] society celebrates, particularly when it comes to writing our resumes. (White, *Atlanta Journal Constitution*, January 5, 2003, R1, R4)

Your e-portfolio can be just what you need to showcase the creativity and individuality that might not show up in your paper resume. It also has the advantage of being viewed online and found by Web search engines. Should you send an unsolicited e-portfolio to potential employees? Use the same judgment you would use when deciding when to send a printed portfolio. A business card with your Web address can be both an advertisement for your online e-portfolio and an invitation to view it.

CONTENTS OF AN E-PORTFOLIO

Basic elements

The basic elements of an e-portfolio are the same as those of a paper portfolio. The major differences are the way they are displayed and how they can be organized and viewed; we'll look at these differences later in this chapter. Figure 5.1 shows several examples of home pages produced by Southern Polytechnic State University students. Notice that each one has a hyperlinked list of the basic elements included in the e-portfolio. You will probably want to add an abbreviated resume page to these lists. Many presenters start with a resume and never develop the e-portfolio beyond this stage. If you do this, the possibilities of your portfolio will never be fully realized, and you'll be left with nothing more than an e-resume. An e-resume, presented as a complete "portfolio," can only be viewed as words with no supporting documents. Your audience may well decide that you haven't yet produced anything worth viewing.

The e-resume

E-resumes have their place, and it's important to know how to put one together. In your e-portfolio, your abbreviated e-resume might include a link to a printable version of your complete resume (see Chapter 8 for a

FIGURE 5.1
Sample Home Pages from Students' E-Portfolios
Source: Used with permission, Tom Burns 2005; used with permission, Sarah Milligan Weldon 2005.

full discussion on how to write resumes). If you need to produce an e-resume as a stand-alone document, it may require different formatting than the resume in your e-portfolio. E-resumes have to be formatted specifically for ease of electronic transfer and storage in databases. In "The Top 10 Things You Need to Know about E-Resumes and Posting Your Resume Online," Katharine Hansen addresses these issues. She focuses on keywords and file formats (.rtf, .txt, .pdf, .html, etc.) that match those required by the potential employer. Here's Hansen's top 10:

1. You absolutely MUST have one [an e-resume]
2. Your e-resume must be loaded with keywords.
3. Your e-resume must be accomplishments-driven.
4. Technically speaking, an e-resume is not too difficult to create.
5. Text-based e-resumes are the ugly ducklings of the resume world, but you *can* dress them up a bit.
6. E-resumes are highly versatile.
7. You must tailor the use of your e-resume to each employer's or job board's instructions.
8. Take advantage of job-board features to protect yourself and get the most out of posting your e-resume on the boards.
9. A few finishing touches can increase your e-resume's effectiveness.
10. Lots of great resources can help with your e-resume (Hansen, "Your E-Resume's File Format Aligns with Its Delivery Method").

Hansen's list is right on target. Spend the time necessary to ensure that your e-resume focuses on your accomplishments, introduces as many keywords as possible, and targets the specific job you're seeking. Take advantage of the many resources available. Think of the e-resume as your way of introducing yourself before presenting examples of your work. Let the e-resume create excitement about the examples of your work that will follow.

You can find excellent information about e-resumes and file formats in Hansen's Web article "Your E-Resume's File Format Aligns with Its Delivery Method." In it Hansen also directs you to Susan Ireland's Web site, www.susanireland.com, for information on posting, e-mailing, and scanning e-resumes. Maria Mallory White lists several helpful Web sites in her article "Web Can Help Job Hunt" (see Appendix A). On Yana Parker's site, damngood.com, "Electronic Resumes," an article by Clara Horvath, California career consultant, gives helpful advice about e-mailing e-resumes: "BOTTOM LINE: Don't send a resume as an attachment unless you're invited or instructed to do so." Why not? Before you send it, Horvath says, you need to know exactly what format the company requires for resumes. Receiving attachments requires extra work. It may be company policy not to open any attachments. Resumes included in the body of an e-mail as ASCII text can be printed, sent to a database, and placed in other formats (Horvath, "Electronic Resumes"). As you prepare to send out your resume, be sure to refer to Chapter 8. And don't forget that an e-resume without supporting documents is not a portfolio; it's just an e-resume.

Examples of your work

Examples of your work are the heart of your e-portfolio, so include only the best. You are better off having three or four excellent examples than several examples of uneven quality. How do you choose those examples? First, review all examples of your work. Chapter 3 gives excellent advice about the selection of representative works and reminds us that these works should illustrate the skills you want to demonstrate to your audience. Be sure to refer to this chapter as you begin selecting examples of your work. In many cases, your instructor will either select the best examples or assist you while providing guidelines. If you are charged with making your own selections, try to achieve consistency for a packaged look that works across both the paper and e-portfolios. Kimball suggests that after listing the artifacts and sorting them by category, you should assess them using a simple score sheet with three to five criteria (Kimball 2003, 50). For instance, when screening works for an artist's portfolio, you might use the following criteria: a good drawing has a complete range of values from light to dark, has a definite focal point, and uses design principles to create a unified composition. Rank each example according to each criterion, and then choose only the works that score the highest.

Reflections on your work

Since the purpose of your presentation portfolio is to demonstrate your proficiency through examples of your work, context statements can be extremely helpful to the viewer. A good context statement introduces the work, explains the context, outlines the developmental process, and assesses the work (Kimball 2003, 22). In other words, the context statement explains what you've done, why you've done it, how you did it, and why you believe it is successful. As you work through this book, you'll receive instruction and guidance on putting together context statements for your portfolio entries (see Chapter 4). Again, the design of the context statements in your e-portfolio should be similar to the design of those in your paper portfolio for consistency and recognition.

PLANNING E-PORTFOLIOS

Now that you know what you want to include in your e-portfolio and have provided contextual information, you're ready to plan how to put this portfolio together. It is now time to do the following:

- Create a written plan including a list of works, file names, and the structure of the portfolio.
- Sketch ideas for a page design template including the title, contents, navigation, graphics, and table cells indicated by dotted lines.
- Create a site map showing links between each page. (Kimball 2003, 58–61)

Make the list of works first, including all the file names. How do you want to organize the files? Do you want to use a unifying theme or metaphor to provide structure? Refer back to Chapter 3 for guidance on understanding themes and metaphors. This information can be especially useful for e-portfolios, since metaphors are inherently visual.

Now you're ready to sketch some ideas for the design of a template—a "page" that contains information repeated on all pages of the portfolio. The template will actually be designed and created later; right now, you're just trying to develop ideas about how you want your e-portfolio to look. These ideas should include graphic elements, navigation bars, titles, and spaces where different entries in your portfolio will be viewed. Don't worry too much about technical issues at this point; just sketch where you think you want these items to appear onscreen. We'll create a site map later on in the chapter.

EXERCISE 5.1 SKETCH A DESIGN FOR YOUR HOME PAGE

List the items that you plan to include on the home page. Decide on a theme, metaphor, or look. Then sketch your idea of what you want the home page to look like. See Assignment 1.

DESIGNING E-PORTFOLIOS

All visual communication has an inherent design, that is, relationships governing the way the parts are put together. Not all of these designs are well considered. Gestalt theory provides laws that govern how we interpret what we see; when visual communication is organized according to these laws, the resulting design allows you to easily scan, read, and interpret the information in the most efficient way with the least distraction. These Gestalt laws are translated into principles of design; these principles should guide you in planning your e-portfolio.

Principles of design

Learning how to apply design principles is the next critical step. These principles are derived from perceptual laws defined by Gestalt psychologists (Berryman 1990, 8; Ryan 1997, 33). You'll formalize the ideas you've

FIGURE 5.2
Proximity Causes Us to See the Items Closest to Each Other as a Group

already sketched out by applying these principles to make a well-designed template that will be used by all the pages of the site. There are many different names for design principles, but Robin Williams, author of *The Non-Designer's Design Book* and, with John Tollett, of *The Non-Designer's Web Book*, has conveniently consolidated the major Gestalt principles into four convenient categories—proximity, repetition, alignment, and contrast (Williams 2003, 14). These are the categories used in this chapter. As you learn about these principles, consider applying them not only to the design of your e-portfolio but also to all the printed documents within the port-folio. It might be helpful to remember the principles of design by an acronym—PRAC—as in "PRACtice makes perfect." If you practice applying these simple principles your e-portfolio will be efficient, well organized, and easy to read and navigate. The next few sections will explain how to apply these design principles.

Proximity.   The Gestalt Law of Proximity states that when two items appear close to each other, we tend to see them as belonging to the same group, as Figure 5.2 shows (Berryman 1990, 8). If we apply this principle to information, we should carefully organize information so that related items are grouped together and un-related items are visually separate. Using the same font style, color, and size for related information empha-sizes the relationship. Repetition reinforces proximity. This means that if two unrelated items look the same (e.g., have the same font, color, and size), then they will be seen as related even if they aren't. When two ob-jects are close together, we see them as a group and immediately assume that they have something in com-mon, in contrast to objects outside the group. Knowing this can greatly improve visual organization; group like information and objects together; separate items that don't relate. Adding extra "white" space can be a very effective way to separate unlike items or groups.

Repetition.   The design principle derived from the Gestalt Law of Similarity, the repetition of colors, graph-ics, fonts, and other elements, states that when items look similar to us, we see them as part of a single group (Berryman 1990, 9; Ryan, 34). Grouping by repetition can include, among other things, using the same fonts for heads and subheads, using the same color scheme throughout your e-portfolio, and repeating the same layout on each page (Williams 2003, 43).

Kimball suggests that sans serif fonts are more casual and serif fonts all more formal (see Figure 5.3). He does not distinguish between the legibility of the two onscreen, and he suggests that the most common fonts are Times New Roman, Courier, and Arial (Kimball 2003, 75). Williams and Tollett, however, suggest that when choosing fonts, it's generally agreed that the sans serif, or "flagless," fonts create text that is easier to read on a monitor screen (and consequently online) than the serif fonts (Williams and Tollett 1998, 214). The reason is that sans serif fonts are vertically constructed with even strokes, and they are more easily constructed with tiny squares, or pixels, than serif fonts, which are created with both thick and thin strokes, frequently on the diagonal. Commonly used sans serif fonts are Arial, Helvetica, and Verdana. Mario Sanchez, who publishes *The*

FIGURE 5.3
Sans Serif Letter (Left) and Serif
Letter (Right)

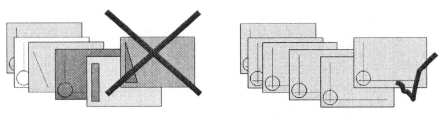

FIGURE 5.4
Repeat Your Design on Each Page for Unity in Your Site

Internet Digest, a great Web resource, suggests that usability tests find Verdana to be the best sans serif font for Web use; Sanchez also says that Georgia is the best serif font to use (www.theinternetdigest.com). Be aware that the default font for Netscape Composer, the popular free "What You See Is What You Get" (WYSIWYG) Web editor, is Times New Roman, a serif font. This means that you must make an effort to change the default font to promote good design. For headlines, subheads, and other groups of short words in large font sizes, you can use the serif fonts if you choose. Avoid using all capital letters, since they are harder to read (Williams and Tollett 1998, 215).

Repeating thematic graphics throughout can do a great job of unifying the e-portfolio. By contrast, changing the design from page to page destroys unity and confuses the viewer (see Figure 5.4). Brian D.'s creative resume (see Figure 5.5) sets the theme for his e-portfolio: "The Tech Writer Marathon: I'll go the extra mile."

FIGURE 5.5
Student Brian D. Uses an "Extra Mile" Theme
Source: Used with permission, Brian Dooley 2005.

The tech writer marathon

Technical writing is like a marathon with many important milestones to achieve in any given project. And it's always a race against the clock. When the starting gun goes off, a technical writer needs to be ready to get the project going and see it through to the end. This is no simple task. Training, skill, and talent will all be tested. A good technical writer will get the job done right and get it done on time. A good technical writer will go the extra mile.

Course map

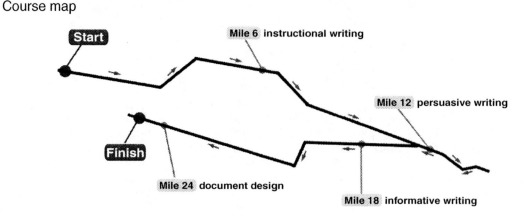

FIGURE 5.6
The "Extra Mile" Theme Is Extended to the Design of the Site Map
Source: Used with permission, Brian Dooley, 2005.

Notice how he extends the theme's metaphor by designing a site map that looks like a marathon course map (see Figure 5.6). Different sections of this portfolio appear as mile markers on the course. Finally, Brian D. uses larger graphics of the mile markers to introduce different sections of the portfolio (see Figure 5.7). The images provide excellent unifiers while adding visual interest. Repeating and extending the portfolio theme, his graphics not only highlight his clever organization scheme but also give the viewer more than a glimpse of his creativity.

Alignment.   Alignment, the principle based on the Gestalt Law of Continuation, states that when two or more items appear lined up, either one below the other or side by side, the eye tends to search for additional items by continuing to move in that direction (Berryman 1990, 9; Ryan 1997, 35). This is why we align type to the left, one line under the other. Type aligned this way is easy to scan and can be read quickly (see Figure 5.8). Many content lists are columns of text aligned left to promote easy scanning. An alternative is horizontally aligning the list of contents on a menu bar at the top.

FIGURE 5.7
Mile Markers on Each Page
Unify Brian D.'s E-Portfolio
Source: Used with permission, Brian Dooley, 2005.

FIGURE 5.8
Type Aligned On the Left Is Easy
to Scan and Fast to Read

- Align items for quicker scanning
- Align items for easier reading
- Easier to group related items

Type centered on a central axis creates unnecessary eye movement, is much harder to read, and is difficult to scan (see Figure 5.9). This type of alignment occurs primarily in formal invitations; you should leave it exclusively for those types of printed documents. Type aligned to the right also requires unnecessary eye movement as the eye searches for the first word in each sentence or phrase. This is why right alignment is generally used only for short headlines and to introduce a surprise element into your page design. Alignment issues affect more than the text of your e-portfolio; they also affect the placement of headlines and subheads. For quick scanning of a Web site, it is very useful to align headlines and subheads to the left. If you align captions either directly under photos or close to the right side, you will also help your viewer see the photo and the caption as a unit. Use the same rules of alignment when you place series of objects. Scattering an arrangement of objects with no apparent organization, like those in Figure 5.10, causes visual confusion. Line up these objects either vertically or horizontally, like those in Figure 5.11.

Contrast.   Contrast tells your viewer what to see first. The principle of contrast comes from the Gestalt principle of figure/ground, which states that if there is sufficient contrast, your viewer can separate the figure, the object to be viewed, from the ground or background, or its surrounding context (Berryman 1990, 9; Williams 2003, 53). If you don't have enough visual contrast, then the viewer can't find a visual starting point—a focal point (see Figure 5.12). Consequently, the viewer has to decide where to start in order to begin making sense of the information. Contrast is what identifies the starting place, and this contrast establishes a visual hierarchy of information. There must be enough contrast between the title or headline and the subheads and text so that the viewer can recognize the difference and start with the item of greatest contrast (see Figure 5.13).

You can create this contrast with type size, style, color, and/or weight. In order to begin categorizing, or making sense of visual data, the eye looks for the largest, darkest item first (Arnheim 1969, 67). This should be the headline. Then it looks for the next largest, darkest item to link them together; this activity should link the headline to related subheads. These subheads should be only slightly smaller than the headline or title. Only after visual relationships are created between the head and subheads should the background—the text— be considered or noticed. And in the process of reading the text, the brain seeks links between the subhead and the text that follows. If the relationship isn't clear, the brain can't process the data efficiently.

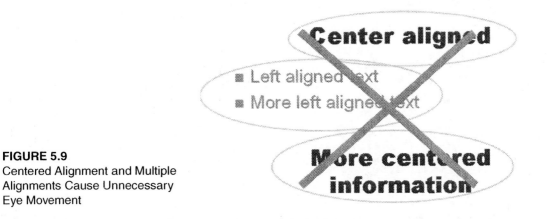

FIGURE 5.9
Centered Alignment and Multiple
Alignments Cause Unnecessary
Eye Movement

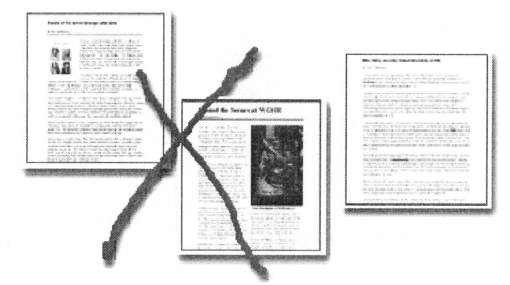

FIGURE 5.10
Unorganized Objects Cause Visual Confusion

FIGURE 5.11
Align Objects for Easier Searching and Scanning

Headline

Subhead
Text is same size as subhead and head, resulting in lack of contrast. There is no visual hierarchy here.

FIGURE 5.12
Insufficient Contrast

Headline

Subhead
Text is the smallest item. This creates a visual hierarchy and allows the viewer to scan the material quickly. The subhead is closer to the text than to the headline; this helps identify the subhead with the text.

FIGURE 5.13
Contrast Creates a Visual Hierarchy

Unity.   When all these principles work together, unity, or completeness, is achieved and the viewer can process data efficiently and quickly. This unity is defined by the Gestalt Law of Equilibrium, sometimes called the Law of Simplicity or Good Configuration (Berryman 1990, 9; Ryan 1997, 35). It states that there is an apparent order and efficient organization. In this case, each Gestalt principle is used to reinforce the other. Bulleted items contrast sharply with text that is aligned farther to the left. The bullets, aligned underneath each other and appearing as repeated dots in close proximity, provide great contrast with regular text and command attention by their contrast (see Figure 5.14). Text, aligned and spaced in proximity as a closely knit block, contrasts greatly in size with the headline and subheads. The small size and relatively light appearance of the text cause it to recede into the background as context for the head and subheads.

The information that follows will be bulleted:

- The text is aligned so that the viewer views the information as a list.
- The dots to the left are repeated, which causes the text to be perceived as related.
- The dots are close together, further strengthening the relationship.
- The physical arrangement of the items contrasts with the text above and below, creating contrast that differentiates the bulleted list.

FIGURE 5.14
Bullets Provide Contrast with Text and Visually Organize Information

Other Gestalt issues

The Gestalt Law of Isomorphic Correspondence states that certain visual images can trigger associations (Berryman 1990, 9). Sometimes these associations are unpleasant, unproductive, or distracting. For this reason, images have to be chosen carefully; generic images are almost always more desirable than specific ones since they are less likely to cause association with particular thoughts or memories (see Figure 5.15).

FIGURE 5.15
Too Much Detail (Left); Just
Enough (Right).
Source: Clip art from Microsoft Office; image on the right altered slightly in Photoshop

EXERCISE 5.2 APPLYING GESTALT PRINCIPLES OF DESIGN

Using the Gestalt principles of design discussed in this chapter, analyze three portfolios that you find on the Internet that either use or abuse these principles. Write up your analysis (approximately one paragraph per portfolio).

CREATING TEMPLATES

After gathering all documents and artifacts into a working portfolio and determining what your e-portfolio needs to show and who will see it, you will create the template. This template can be as simple or complex, as plain or fancy, as you choose. Just remember that your template is basically the framework—a consistent design and structure applied throughout your e-portfolio; keep the design clean and functional. Now you can take out your earlier sketches and make a more detailed drawing of how you want your home page to look. Using templates will ensure that the appearance of your e-portfolio remains constant (an example of the principle of repetition); it will also save a lot of time. The templates used in illustrations in this chapter are very simple ones designed for those just beginning to explore e-portfolios. Once you feel comfortable with design and layout and how to translate these into an electronic format, you can be as creative as you want in developing your own template.

Knowing what will be included in your e-portfolio is the first step in creating a table of contents. The table of contents should appear on the first page, or home page, of the e-portfolio as a navigation bar or list of links to other pages or sections (see Figure 5.16). (You may choose to begin your e-portfolio with an introductory screen that directs the viewer to the table of contents page.) A possible list of sections with links might include Resume (abbreviated), Biography (also abbreviated), and Projects and Reflections on Projects. These links will appear on all pages, along with a Return to Home Page link.

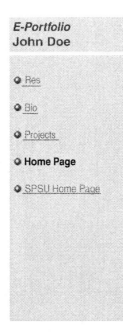

FIGURE 5.16
A Simple Navigation Bar from a
Student's E-Portfolio

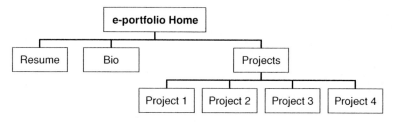

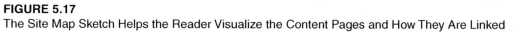

FIGURE 5.17
The Site Map Sketch Helps the Reader Visualize the Content Pages and How They Are Linked

Designing and placing these navigation links is a major part of your e-portfolio design. Your home page should identify you while including critical information such as contact information and a brief overview of the purpose of the e-portfolio. For personal security, consider limiting the contact information to your e-mail address. Choose a color scheme of no more than three colors, or perhaps use only black, white, and gray, but whatever you choose, carry the scheme throughout the portfolio. This consistency, or repetition, gives a feeling of unity to your site. Avoid being "cutesy," and don't try to entertain viewers with dancing figures, flashing lights, graphics, or other extraneous information. Above all, remember that your e-portfolio is a professional presentation. Before committing anything to HyperText Markup Language (HTML) code, sketch several ideas for a template design (see Figure 5.17) and decide where to put the navigation bar and where to locate blocks of information. This helps to determine which elements will appear on every page and which ones will change from one page to another. Those occurring on every page will be saved on the template exactly as they will appear, and blocks of space on the template will be reserved for those elements that will change. On the computer, create a folder that will hold all the files needed for the project. If you create folders for documents and graphics within this main folder, inserting images and creating links will be much easier. This is also where you will work out a plan for linking pages by creating a site map of the e-portfolio. Doing this will save hours of unnecessary computer work.

EXERCISE 5.3 CREATING A SITE MAP OF YOUR E-PORTFOLIO

Create a site map of your e-portfolio by sketching out each page and links between the pages, as shown in Figure 5.17. See Assignment 3.

Getting your template into HTML format

Getting your e-portfolio template into HTML format can be simple or complicated, depending on how elaborate your design is. If you're a beginner, you should probably use a WYSIWIG HTML editor—a program or application that writes the HTML code for you while you design its appearance in a desktop publishing environment. When writing straight HTML code, you can't see what the page will look like until you open the coded page in an Internet browser; by contrast, using a WYSIWYG editor allows you to see what the Web page will look like while you create it. And there's an additional advantage: you don't need to know how to write HTML code to use a WYSIWIG editor.

Microsoft FrontPage, Adobe PageMill, and Macromedia Dreamweaver are popular commercial WYSIWIG editors. With its browser, Netscape provides an excellent free WYSIWIG editor, Composer; both the browser and the editor can be downloaded for free from Netscape's Web site. You can download a very similar open source WYSIWYG HTML editor from the Web for free—NVU (pronounced "n-view"0)—from NVU's Web site (see Appendix A for the address).

Web site design becomes more complicated when you include frames on the Web pages; if you're a first-time designer, you will find the process much less frustrating if you keep pages simple and frame-free. Using tables to contain blocks of information may be the simplest way of keeping text and graphics in place. Planning a layout using a table is much like planning a layout on a grid. The cells in the illustration (Figure 5.18) are part of a table; the illustration itself is a simple Web home page created in Composer and viewed in the Internet Explorer browser. The cells of the table can contain information, links, background color, and even graphics. The boundaries of the cells keep the information positioned in the same place; otherwise, there would be no control over how and where the elements appeared on the Web page. The Web page in Figure 5.18 looks like Figure 5.1 within the Composer HTML editor.

There are many resources for using tables in Composer; a few of them are listed at the end of the chapter for your convenience. The resources for Composer work equally well for NVU. If you use a table format, you must first decide how many rows and columns you need. You can determine this by sketching on paper the elements you want to repeat on all pages, as well as those that will change from one page to another (see Figure 5.20). Figure 5.19 shows how the template sketch is translated into an HTML table within the HTML editor. The cells in the large gray area have been merged, or made into one large cell. This allows links to be typed with different spacing—which wouldn't be possible if confined to different rows. Figure 5.20 shows what this table looks like when viewed with the Internet Explorer browser. Even so, creating a template and then saving a copy of it for each page will make the creation of your e-portfolio much easier.

Although this chapter doesn't attempt to tackle the use of XHTML (Extensible HyperText Markup Language) to create e-portfolios, those of you who are proficient in HTML should consider XHTML's possibilities. This next-generation language, a type of XML (Extensible Markup Language), extends the uses of HTML by allowing data to be accessed, stored, and manipulated more easily and in many more ways. There are many Web sites that are excellent resources if you're interested in learning more about XHTML. It is important to

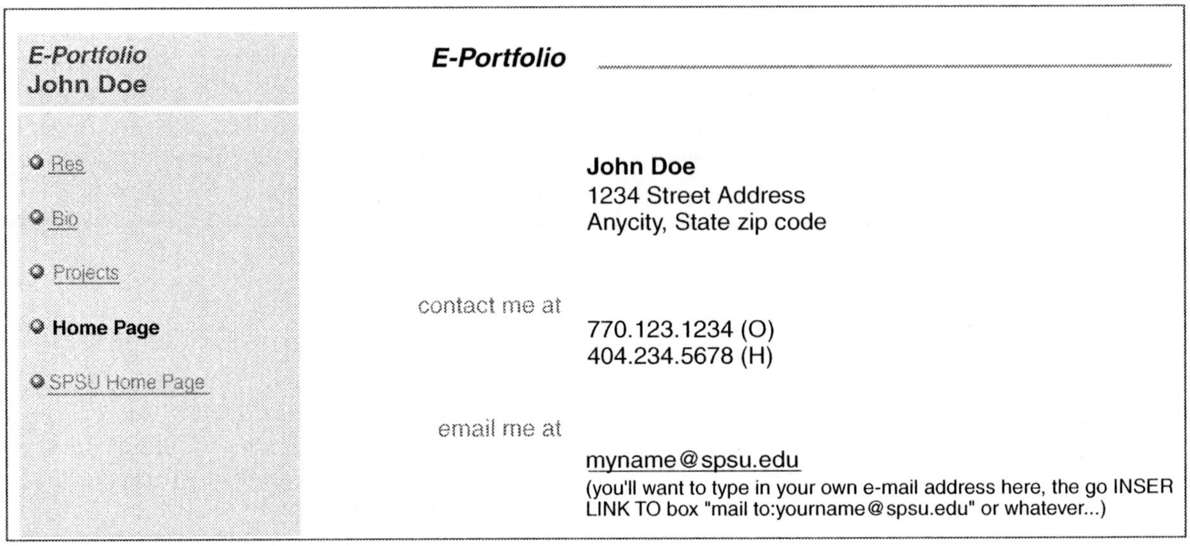

FIGURE 5.18
A Simple HTML Home Page

E-Portfolio John Doe	e-portfolio	
● Resume ● Bio ● Projects ● **Home Page** ● SPSU Home Page		**John Doe** 1234 Street Address Any city, State zip code
	contact me at	770.123.1234 (O) 404.234.5678 (H)
	email me at	mailto:yourname@spsu.edu (you'll want to type in your own email address here LINK TO box "mail to:yourname@spsu.edu" or what

FIGURE 5.19
A Simple HTML Home Page Viewed within the Netscape Composer HTML Editor

recognize that the technology in this area is constantly changing. As a professional communicator, you will need to stay current on these technological advances and adjust accordingly. The technical advice presented in this chapter is meant as a starting place to give you a sense of process and structure.

Getting your template into PowerPoint format

If you don't want to use HTML and are familiar with PowerPoint, then creating a PowerPoint e-portfolio is an option. Using PowerPoint won't make the presentation better; it may make it easier to create. In the long run, it would benefit you to spend the extra time becoming familiar with an HTML editor. If, however, you choose to use PowerPoint, then before you begin, create a folder to save the presentation. Within this folder, create another folder for all images and graphics. Keeping your files together this way will make your work much easier. Next, create a template that will become the Master Slide. Your design challenge in PowerPoint is to override the default settings of the templates or to create your own. Settling for a ready-made template is a red flag, alerting the viewer that you are an amateur user of the application.

Nav Bar John Doe	E-Portfolio of John Doe	
Res Bio Projects Home		John Doe 1234 Street Address Any city, State Zip 770.123.1234 (O)
	Contact me at:	404.234.5678(H)
	e-mail me at:	myname@spsu.edu

FIGURE 5.20
A Template Sketch Helps You Determine the Number of Rows and Columns Your Table Needs

It is extremely important for you to create a unique personal template that reflects your personality and skills. Restraint is especially important in PowerPoint; it is not only tempting but all too easy to add animation, sounds, and special effects. Pass these up, and concentrate instead on creating a clean professional presentation that focuses on your work rather than your cleverness. Chapter 3 deals with this issue and discusses the use of themes and metaphors in portfolio design. Once you design the PowerPoint template and save it as the Master Slide, you can customize each page with the proper information. You can easily add links to graphics and supporting documents. For instance, Figure 5.21 shows you how to add an e-mail link to the PowerPoint slide. Adding links to other pages is just as easy. Go to INSERT>HYPERLINK...> and click on PLACE IN THIS DOCUMENT, then browse to the page you want to link to and click on it. Adding images to a slide is even easier; just click on INSERT>PICTURE>FROM FILE and browse to find the file, then highlight the file and click OK. The file will be inserted in the document.

ADDING GRAPHICS

Graphics obviously play a vital role in visual communication, and your choice of graphics is important in determining the quality of your e-portfolio. While Kimball (2003, 76) suggests using clip art but warns against violating copyright laws (see Chapter 6 for a discussion of copyright law as it pertains to portfolios), you would do well to avoid clip art entirely if possible. Stock images like clip art suffer from overuse, and using such images in an e-portfolio may suggest that you have taken the easy way out rather than selecting or creating a unique graphic designed for a specific purpose.

A viewer pays more attention to something he or she has never seen than to something he or she has already seen several times. We also know that resumes in general are likely to receive only cursory glances. So, using a piece of clip art simply cues the viewer to expect the same old thing rather than something unique and unusual. If you avoid clip art, then you must learn to create your own unique graphics. These graphics will support the theme or metaphor of your e-portfolio. The signposts in Figure 5.22 are good examples of graphics modified to support the metaphor of a student portfolio. It's all too easy to find a neat graphic and then try to build your portfolio around it; the problem is that your portfolio will never catch up with your graphic but will always be subservient to it rather than the reverse. Remember, the e-portfolio's purpose is to showcase your work, not to show off a neat graphic. Having said that, let's discuss how to create and use graphics.

FIGURE 5.21
Adding an E-Mail Link to a PowerPoint Slide

FIGURE 5.22
Nanette's Customized Graphics

Creating graphics

Using a scanner or digital camera to create unique graphics is not only relatively easy but also provides an opportunity to be creative. The main challenge in using both a scanner and a digital camera is preparing the resulting digital image for electronic publication. Preparation includes selecting the correct resolution, the proper file format, and the appropriate size for digital display.

Using the correct graphic file format. Online graphics should be in a file format designed specifically for viewing onscreen. Keeping it simple, the most familiar formats are Joint Photographic Experts Group (JPG) and the Graphic Interchange Format (GIF). The JPG file format compresses the file, reducing its size and losing some of the information. It is most useful for photographic images, being able to display millions of colors. The GIF format, on the other hand, is best for flat colors, since it uses a very limited color palette. Portable Networks Graphics (PNG) is another, less familiar, graphics file format designed specifically for online graphics; it compresses the image to reduce the file's size while retaining all of its information. Graphics to be displayed onscreen will usually be saved in one of these file formats.

Understanding graphic size and resolution. On the electronic screen, graphics are displayed at a standard resolution of 72 pixels per inch (ppi). The computer screen is composed of pixels, and although PCs and Macs actually have different screen resolutions (PCs, 96 ppi; Macs, 72 ppi), it's commonly accepted that 72 ppi is the standard means for ignoring the differences when dealing with graphics (some choose to use 75 ppi—not much difference, but the math works better). Scanners also use pixels per inch to describe image output. Dots per inch (dpi) refers to the dots that a printer prints per linear inch. When creating an e-portfolio, you are concerned mostly with pixels per inch, since these determine how the material appears onscreen. Without a complete understanding of this principle, many graphics posted online appear unnecessarily large, and the viewer has to scroll to view the whole image. Since the monitor screen is effectively capable of displaying only 72 ppi, you can see in Figure 5.23 that a 72×72 pixel image takes up 1 square inch, while a 150×150 pixel image takes up about 2 square inches and a 300×300 pixel image takes up a whopping 4 square inches!

FIGURE 5.23
The Onscreen Appearance of the Same Image, 1 Inch Square, with Three Different
Resolutions—72, 150, and 300 ppi

Changing resolution.   Knowing how pixels are displayed, you must be sure that the images used in your
e-portfolio are saved at the proper resolution for electronic display—72 ppi. It's best to scan and create your
image files at the highest resolution possible and then create a copy at low resolution for your e-portfolio.
Changing an image's resolution is easy and can be done in several digital imaging applications. In Photoshop,
you can reduce the resolution of an image easily by pulling up the "Image Size" dialog box under Image on
the menu bar, as shown in Figure 5.24, and typing in the new lower resolution.

 Now note the changes that are registered under Pixel Dimensions when the resolution is changed to
72 ppi (see Figure 5.25). Notice that when the Constrain Proportion box is checked, brackets appear to the

FIGURE 5.24
Clicking on Image>Image Size
in Photoshop Reduces the
Resolution of Graphic Files to
72 ppi for Onscreen Display

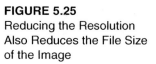

FIGURE 5.25
Reducing the Resolution
Also Reduces the File Size
of the Image

right of both the width and height boxes, locking these dimensions together so that if one dimension is changed, the other changes proportionately. As you can see, changing the resolution from high to low is easy. If you are working with a scanned image, always make the initial scan at 300 ppi and archive this image. To use this file at a lower resolution, first make a copy of the file and then change the resolution of the copy. Notice that these instructions don't include changing the resolution from low to high; this is something you should avoid, since the image quality will deteriorate noticeably.

Changing the graphic size. The document size of a graphic can be changed just as easily as the resolution and in much the same way. By changing the desired dimensions (either pixels or inches) in the Image Size dialog box and clicking OK, you can automatically resize the image (see Figures 5.26 and 5.27). If you don't know what pixel dimensions the graphic should be, figure out the size of the image in inches in width or

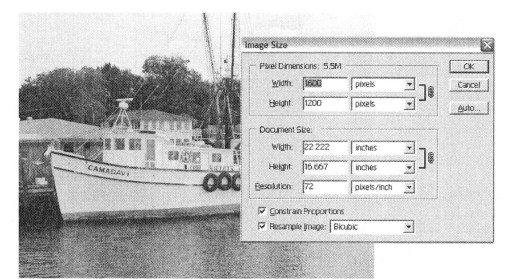

FIGURE 5.26
Original Document Size

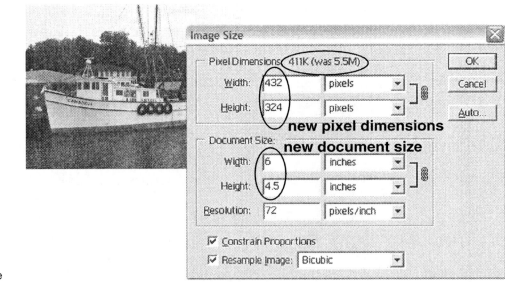

FIGURE 5.27
New Document Size

height. Determining only one dimension is necessary as long as you check the Constrain Proportions box. When this box is checked, changing one dimension will automatically change the other proportionately. Changing the pixel dimensions will also change the document size automatically.

Inserting graphics into the HTML document

In Composer, NVU, and other WYSIWIG editors, inserting a graphic file is almost as simple as inserting a graphic file in PowerPoint. Click in the cell where the image is to be inserted, then go to INSERT>IMAGE…>CHOOSE FILE>, highlight the image, and click OPEN. Fill in the Alternate Text dialog box by briefly describing the image and click OK. The image will appear in the cell, as shown in Figure 5.28.

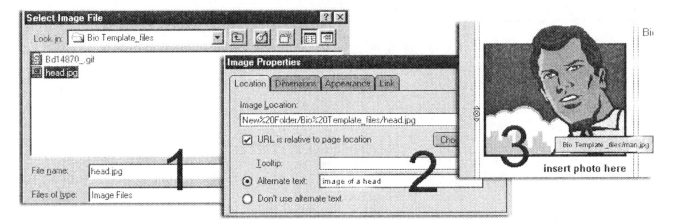

FIGURE 5.28
Inserting a Graphic File Is as Simple as 1-2-3

USING PORTABLE DOCUMENT FORMAT (PDF) FILES

Entries in your e-portfolio's table of contents will be linked to the documents and graphics you've created so that your audience can see them. These graphics and documents should be in Portable Document Format (PDF). Then your viewers can see and even download your files exactly as you created them without having to own the software you used to create them. These PDF files can be viewed by anyone who has Adobe Reader on his or her computer; the Reader is free from Adobe.com. It's a good idea to alert your viewers that Reader is necessary for viewing PDFs and to share with them the address of—or, better still, a link to—the site where it can be downloaded for free. Creating PDF files, on the other hand, requires either purchasing the complete Adobe Acrobat application or other software that creates PDF files. PDF995, freeware, is available for download from PDF995.com. When creating PDF files, it's important to keep the file size as small as possible. You can do this by creating black/white or grayscale files whenever color isn't necessary and by keeping the resolution as low as possible. Usually 72 ppi is sufficient resolution for clear display and reading onscreen. Don't expect to get photographic-quality prints at this resolution. If you need a good-quality print of the document, save the document at 300 ppi if possible.

Software that creates PDF files installs a driver on your computer that allows it to "print to" this driver, much like printing to a copy machine. The computer then "copies" the electronic document file and saves it as a PDF file. To "save as" a PDF file, simply choose the PDF file format on the pull-down menu. If this is not a choice, once the appropriate software is installed, you can "print to" PDF. When "PRINT" is selected, the dialog box shows which printer will be used. Use the pull-down menu to select "PDF995," "Adobe Distiller," "PDF Writer," or another PDF creator (see Figure 5.29). Then select a name and location for your new PDF file (see Figure 5.30).

That's all there is to saving a computer file as a PDF file. If you have a paper document without the accompanying file, you must scan the file to save it in PDF format. This is no more difficult than scanning a file to any other format. However, the appropriate PDF software must be installed to scan as a PDF file. You may

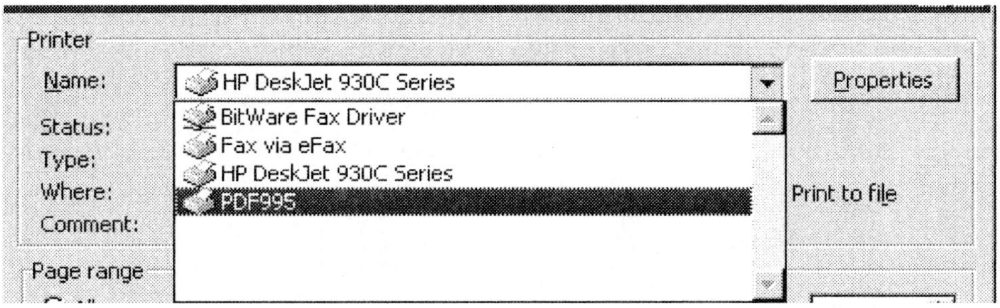

FIGURE 5.29
To Create a PDF File, Select the PDF Software in the Print Dialog Box

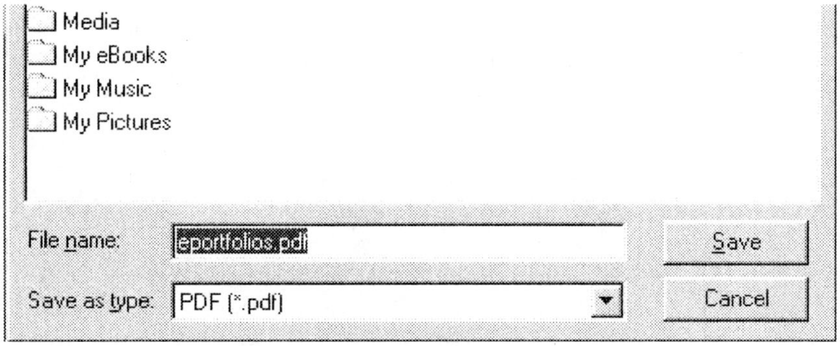

FIGURE 5.30
In the Save As Dialog Box,
Choose a Name and Location to
Save the File

choose to save your documents as HTML files rather than PDFs, so it's important to know that saving documents as HTML files often changes the formatting. Consequently, some documents saved as HTML files don't look the same as they did in the original file format. Saving as an HTML file from a desktop publishing application can also add additional—and often unnecessary—HTML code that can be troublesome to change. The advantages of PDF files are that the formatting doesn't change and your viewers will see the document exactly as you created it, even if they don't have the application you used to create it. Saving as HTML may be easier, but in the long run, creating PDF documents may look more professional.

SAVING YOUR E-PORTFOLIO IN HTML FORMAT

Determining the purpose of your e-portfolio and its audience will dictate where you save the HTML and PDF files. If you want an unlimited audience, then you should probably publish your e-portfolio as a Web page. If your e-portfolio is intended for a limited audience, then your HTML and PDF files can be saved to a CD or DVD that can be mailed. Or if your files are small enough, you can create a zip file of your e-portfolio and e-mail it. Regardless of the final format, files generated in an HTML editor like Composer will be saved as HTML, and files created in PowerPoint can be saved as HTML with little effort.

ADDING AUTO-RUN TO YOUR E-PORTFOLIO

If you transfer your e-portfolio to a CD, you may want to include an "auto-run" feature. This feature causes your CD to open automatically on a computer. If your e-portfolio is on an auto-run CD, it will open automatically for viewers. This is a simple matter of adding additional files to the root folder on your CD. There are many applications you can buy that will create auto-run CDs, but you can just as easily create your own by downloading the files and then customizing them slightly. Softwarepatch.com offers a free download of an auto-run zip file; just unzip it and, following the "readme" instructions carefully, place the three files—autorun.inf, autorun.bat, and index.html—in your root folder. Before you place the files on the CD, open the autorun.bat file in Notepad (don't double-click; use "open file") to make any minor changes in the "comment" section; in the second line of the code, change the name of the file to be opened (if other than index.html), and then save the file as autorun.bat. You've created a CD that opens your e-portfolio automatically!

ADDING AUDIO TO YOUR E-PORTFOLIO

As you probably already know, you can add audio files to PowerPoint slides, but you may not know that you can add to HTML files audio files that play automatically. You can use this feature to give a brief explanation of a e-portfolio entry or to welcome the viewer to your e-portfolio. Using audio files is another easy way to personalize your e-portfolio. To add audio files, you must first create them. Using the Sound Recorder feature on your computer, record your message, keeping it short, and save it as a WAV file. Always play the file back to make sure that the sound quality is good. If not, record it again. Next, add HTML code to link the audio file to the document. DevX.com provides clear instructions and HTML code for adding an embed tag. By inserting this code into your HTML document, you can have the audio file open automatically and play as the Web page opens. Not only can you show viewers what you can do, you can also tell them!

UPLOADING YOUR E-PORTFOLIO

If you want to publish your e-portfolio on the Web, you must place it on a host server so that others can access it. At a college or university, the easiest and least expensive way to publish an e-portfolio is to place the files on the school server. If your school allows this, then the files will be uploaded using either the file transfer protocol (FTP) or perhaps even a drag-and-drop method. If this is not available, then you have two other options: find a free Web site host or pay for a hosting service. Ultimately, you will probably want your own Internet service provider (ISP); free use of college or university server space usually ends with graduation, and that's when your e-portfolio will play its most important role. Free hosts usually require that your Web site include an ad for the service and may include other ads as well. The charges for a commercial service will vary, depending on the services obtained. These services may include Web design tools, design service, customer support, storage, and e-mail service. See Appendix A for sites that offer step-by-step instructions for preparing your Web site and uploading it.

If your e-portfolio will be hosted on a server other than your college or university server and if you want an address for your site, then you need to obtain a domain name. Potential viewers will then be able to access your site by typing the domain name into their Internet browser address bars. Domain names can be obtained from various domain name registrars; these registrars will then register your name with the InterNIC Registration Service. Fees are relatively low and are based on how long you want the registration to last. Once your domain name is registered with an ISP, your e-portfolio is ready to be uploaded. This will be done with FTP. You can download many FTP programs for free from the Internet. Using an FTP program is relatively simple; the program allows you to log into a remote server and then simply transfer files from your home computer to the host server.

GETTING SEARCH ENGINES TO FIND YOUR E-PORTFOLIO

After you've gone to all the trouble of designing and putting together an e-portfolio, securing an ISP, and registering a domain name, you should make sure that search engines find the site. You can register your site for free with all the major search engines, but to ensure that your site is registered with as many search engines as possible and to make this process as easy as possible, you may want to pay a service to handle this. See Appendix A for addresses that provide these services. There are also many search engine optimization (SEO) tools available to help you prepare the site for search engine recognition if you want to undertake the process yourself.

After the e-portfolio is uploaded and registered with search engines, you should take a well-earned rest while your e-portfolio does the work you designed it to do—sharing with potential employers your organizational skills and creativity!

SUMMARY

E-portfolios, once a novelty, are quickly becoming a necessity. To be noticed in the marketplace today, your portfolio must be distributed widely. To be effective, it must display your work to the best advantage. It's hard to find a better option than the e-portfolio for publishing your best works widely, even worldwide, in an instant at little or no cost. With some effort and a bit of creativity, you can publish your work around the globe and create a world of opportunity for yourself.

ASSIGNMENTS

Assignment 1: *Sketch a Design of Your Home Page*

List the information that you want to include on your home page.

Name_____

Contact information_____

_____

Contents of e-portfolio
* E-resume
* Examples of your work (list)

_____   _____
_____   _____

_____   _____

* Reflections on your work
* Other_____

_____

Theme or metaphor you want to use (if any)_____

Now make a sketch of how you want your home page to look and where you want to locate the elements listed above.

Assignment 2: *Create an Information Hierarchy*

1. Read the following text and then divide the information into three to five categories; these will become the paragraphs of text.
2. Write subheads describing the information in each category.
3. Next, write a short headline describing the information as you plan to present it.
4. Finally, using a word processor, create a one-page document with your information hierarchy.

TEXT

Mary had a little lamb that she had received from her parents for her 12th birthday. She had always wanted a little lamb, but it wasn't until she turned 12 that her parents thought that she was old enough to properly care for a pet all by herself. Mary named her snow-white lamb Agnes after Saint Agnes, a fourth-century martyr. Because Mary took ownership of the lamb when it was very young, so young that Mary had to feed her with a baby bottle, the lamb thought of Mary as its mother, and everywhere that Mary went, the lamb went along behind, even when Mary wasn't expecting company. In fact, one cold morning in January, Mary, having missed the school bus, rushed out of the house tugging on her gloves and snow cap while balancing her books, lunch, and clarinet case on her hip. She frantically dashed down the sidewalk, literally plowing her way through the 12 inches of new snow as she ran the 11 blocks to the local middle school. Mary's parents leave for work right before the bus arrives, and it's Mary's responsibility to get to the bus on time, so when she misses it, she has to walk to school by herself in the snow. Dashing through the front door of the school and into the front office for a "late" slip, Mary failed to notice that she had company the whole way there. She ran down the seventh-grade hall, dashed into her homeroom, and slid into her desk just as her name was called out by the teacher: "Mary! You know it's against the rules to bring pets to Mayberry Middle School!" Just then, the class began to laugh and squeal and jump up and down by their seats. In the midst of the havoc Mary felt a cold nose on her hand, and she looked up to see Agnes dancing around her desk. There were fish in the school aquarium and frogs and bugs in the terrarium, but no one had ever seen a lamb at school before. The teacher loved animals, though, and she allowed everyone to pet Agnes before sending Mary to the principal's office to call her father to come pick up Agnes. And everyone remembered that cold January day as one of the most special days in that seventh-grade year—the day that Mary's little lamb followed her to school.

Assignment 3: *Create a Site Map of Your E-Portfolio*

1. List here the titles of the pages that you think your e-portfolio will need.

2. Using the following rough sketch as a guide, draw a "map" of how your site pages will relate to each other.

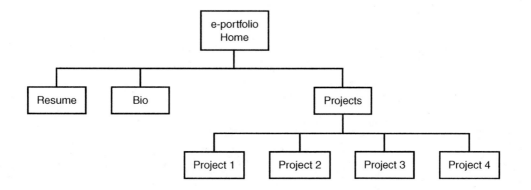

Draw your site map here.

3. From your sketch, determine whether your e-portfolio is linear, hierarchical, or a combination. Write your answer here.

REFERENCES

Arnheim, Rudolf. *Visual Thinking*. Berkeley: University of California Press, 1969.

Berryman, Gregg. *Notes on Graphic Design and Visual Communication*. New York: Crisp, 1990.

Hansen, Katharine. "Your E-Resume's File Format Aligns with Its Delivery Method." http://www.quintcareers.com/e-resume_format.html (accessed April 25, 2005)

———. "The Top 10 Things You Need to Know about E-Resumes and Posting Your Resume Online." http://www.quintcareers.com/e-resumes.html (accessed April 25, 2005).

Horwath, Clara. "Electronic Resumes." http://www.damngood.com/jobseekers/electronic.html (accessed May 10, 2005).

Kimball, Miles A. *The Web Portfolio Guide: Creating Electronic Portfolios for the Web*. New York: Longman, 2003.

Powley, William. "Technical and Scientific Illustrations: From Pen to Computer." *STC Conference Proceedings* 2005. http://www.stc.org/confproceed/1995/PDFs/PG349352 (STC members only; accessed July 26, 2004).

Ryan, Colin. *Exploring Perception*. Pacific Grove, CA: Brooks/Cole and Nelson ITP, 1997.

Sanchez, Mario. "Web-Safe Fonts for Your Site." http://www.accordmarketing.com/tid/archive/websafefonts.html (accessed October 31, 2005).

Villano, Matt. "Hi-Octane Assessment: The Electronic Portfolio Powers-up Student, Educator, and Lifelong Assessment." *Campus Technology* (September 2005): 45–48.

Williams, Robin. *The Non-Designer's Design Book*, 2nd ed. Berkeley, CA: Peachpit Press, 2003.

Williams, Robin and John Tollett. *The Non-Designer's Web Book*, Berkeley, CA: Peachpit Press, 1998.

APPENDIX A: SUGGESTED WEB SITES

http://browser.netscape.com/ns8/ to download Netscape Browser 8.0 with Composer.

http://engineering.dartmouth.edu/~career/handouts/electronic_portfolio.html. Dartmouth's "Developing an Electronic Portfolio" site.

http://www.adobe.com for links to Acrobat Reader and creating PDF files online.

* http://createpdf.adobe.com/?v=AHP to create PDF documents online.

* http://www.adobe.com/products/acrobat/readstep2.html to download Acrobat Reader.

http://www.devx.com/projectcool/Article/20131 for adding audio files to HTML documents.

http://www.eresumes.com/ for samples as well as templates, tips, and tutorials.

http://www.fcs.uga.edu/cs/tutorials/composer/tables.html for help using tables in Composer.

http://www.fluffbucket.com/nsc/tables.htm for help adding tables in Composer.

Maria Mallory White recommends several Web sites in her article "Web Can Help Job Hunt" (*The Atlanta Journal-Constitution*, November 3, 2002, R1):

* http://damngood.com for books, tips, and advice from author Yana Parker.

* http://www.dice.com for posting your resume.

* http://www.execume.com, Gayle Oliver's company site.

http://www.godaddy.com, a popular site for registering new domain names, Web hosting, and so on.

http://www.thesitewizard.com/archive/domainname.shtml for advice on choosing domain names.

http://www.nvu.com to download NVU, a freeware WYSIWYG HTML editor.

http://www.softwarepatch.com/tips/autorun.html for free download of auto-run code.

http://www.w3schools.com/xhtml/for tutorials for using XHTML.

http://www.websitetips.com/xhtml/ for XHTML tutorials, articles, links, and tools.

6 Legal and Ethical Issues Affecting Portfolios

Only one thing is impossible for God: to find any sense in any copyright law on the planet.   Mark Twain (1972)

Introduction

You may find yourself agreeing with Mark Twain's humorous criticism of copyright laws, but, as a professional communicator, you will need to understand intellectual property laws well enough to protect your own work and the intellectual property of your employers, as well as to avoid misusing the intellectual property of others. When creating your portfolio, and later when revising it, you will face a number of thorny ownership issues that you will feel more comfortable addressing if you have some basic information about fair use and public domain. You will also want to know when it is necessary to get permission for work that might be included in your portfolio.

In addition to intellectual property rights, Chapter 6 examines a range of other legal and ethical issues that may occur as you plan, design, and create both your paper and electronic portfolios. We encourage you to develop your own code of ethics to use daily as a professional communicator. You can begin this process by becoming familiar with the published ethical codes of various professional communication associations such as the Society for Technical Communication and the International Webmasters Association. Specifically, Chapter 6 covers the following topics:

* Copyright and intellectual property rights
* Works in the public domain
* Copyrighting your portfolio
* The fair use doctrine and your portfolio
* Work-for-hire doctrine
* Joint authorship
* The Digital Millennium Copyright Act of 1998
* Information liability
* Portfolios and ethical issues

COPYRIGHT AND INTELLECTUAL PROPERTY RIGHTS

Intellectual property is any original work that has commercial value. In its amended form (see the U.S. Copyright Office's Web site, www.copyright.gov for specific amendments), the 1976 Copyright Act remains the central piece of legislation protecting original works of authorship that are fixed, tangible forms of expression. Protection of intellectual property dates back to Article 1, Section 8, of the U.S. Constitution, which grants authors and inventors exclusive rights to their writings and inventions. To be protected by copyright, a work must be both a fixed (in an unchanging form) and a tangible form of expression. Examples of fixed, tangible works include paper documents, digital media, photographs, and videotapes. The terms "fixed" and "tangible" have been interpreted to mean that the original work has a distinct, recognizable form. For example, a logo created for a client is both fixed and tangible because it can be viewed on both the client company's

107

Literary works (i.e. articles, reports, poems, stories) that you may have created for your portfolio

Musical works, including the lyrics

Dramatic works (i.e. any plays that you may have created for your portfolio)

Pantomimes and choreographic works

Pictorial, graphic, sculptured work (i.e. any original photographs or original digital images you may have created for your portfolio)

Motion pictures and other audiovisual works (i.e. original video clips, streaming video, or multimedia projects)

Sound recordings

Architectural works

FIGURE 6.1
Original Works of Authorship Covered Under the 1976 Copyright Law
Source: U.S. Copyright Office, *Copyright Basics* (Circular 1), p. 5.

Web site and its printed correspondence. Knowing how to protect your own intellectual property is important as you create your portfolio. If you have created, for example, an original multimedia project or if you decide to include any original artwork in your portfolio, you should clearly label these original fixed creations as being copyrighted (see the following section on copyright protection). Copyright ownership is a complex issue that will be discussed later in this chapter in the section on work for hire. Original works of authorship are identified in Figure 6.1. It is important to keep in mind that these categories should be viewed broadly. Computer programs, for example, may be registered as literary works, and architectural drawings may be registered as pictures or graphics.

   The 1976 Copyright Act also makes it clear what works cannot be protected by copyright. These works include those listed in Figure 6.2.

WORKS IN THE PUBLIC DOMAIN

The term "public domain" refers to works that are no longer protected by copyright laws or works, such as U.S. government publications, that do not meet copyright requirements. Ideas and facts are not protected by copyright, but tangible, fixed expressions of ideas and facts are. You do not need permission to borrow information from these sources, but you do need to credit the original source.

Works that are not fixed in a tangible form (i.e. improvisational speeches and unrecorded choreographed works)

Ideas, procedures, processes, principles that are not expressed in a tangible form (i.e. not written down)

Familiar symbols or designs

Works comprised entirely of common property without original authorship (i.e. calendars, height and weight charts, lists taken from public documents)

U.S. Government publications

FIGURE 6.2
Works Not Protected by the 1976 Copyright Act
Source: U.S. Copyright Office, *Copyright Basics* (Circular 1), p. 5.

All copyrighted works eventually enter the public domain. For example, Mark Twain's *The Adventures of Huckleberry Finn* (1886) and Lewis Carroll's *Alice's Adventures in Wonderland* (1865) are both in the public domain and have been published in many new editions, some costing only a few dollars. Shakespeare's plays, Da Vinci's "Mona Lisa," and Beethoven's symphonies are in the public domain because copyright laws were not in existence when these works were created. Remember, however, that specific performances and photographs (digital or hard copy) are probably protected by copyright. For example, the book jacket photograph of the "Mona Lisa" that appears on Dan Brown's novel *The Da Vinci Code* is protected by copyright.

Determining whether or not a work is in the public domain is not easy. There is no central database that you can log on to, and the U.S. Copyright Office will not disclose whether or not a work is in the public domain (Fishman 2000, 1/7). You may, however, do a keyword search using your favorite search engine, including as part of your search string the words "public domain." Doing so will produce a range of choices, particularly if you are searching for graphics and/or digital images that are more than likely in the public domain. Another way to avoid copyright issues is to take your own digital photographs.

Public domain and the Internet

As you probably know, a work that is on the Internet is not in the public domain simply because it is on the Internet. Works in the public domain that are on the Internet generally fall into one of the following categories:

* Works without copyright protection
* Works that were in the public domain prior to being posted on the Internet
* Works specifically designed as public domain documents (Fishman 2000, 17/5)

U.S. government publications constitute perhaps the largest body of public domain work that professional communicators might find useful. For example, you may want to include in your portfolio an article on an environmental subject and need photographs for it. You can go to a U.S. government Web site, download whatever graphic you wish as long as it is not already protected by copyright, and use it. Doing so would not be copyright infringement since U.S. government publications are paid for by public tax dollars. Make sure, however, that you give a credit line to the government publication.

EXERCISE 6.1 FINDING A DOCUMENT IN THE PUBLIC DOMAIN

Using your favorite search engine, do a keyword search for a document or graphic in the public domain that pertains to a project in your portfolio.

Print the document or graphic and analyze how you might use it in one of your portfolio pieces.

COPYRIGHTING YOUR PORTFOLIO

As you create your portfolio, you will probably want answers to the following questions:

* Should I copyright my portfolio?
* How do I copyright my work?
* How long does copyright protection last?
* Can I apply for an international copyright?
* What protection does copyrighting my work give me?

Should I copyright my portfolio?

You may think that you should copyright your portfolio since you are working so hard on it and since it reflects your unique talents, skills, and qualifications. Remember, however, that your portfolio is an evolving document, changing as you change and being modified each time you present it to a potential employer, colleague, or classmate. You may, however, want to copyright some of the artifacts in the portfolio if you consider them fixed (unchanging) and if they don't contain someone else's copyrighted work (i.e., a borrowed graphic or sound clip). Another consideration is the format of your portfolio. If you decide to Web host it or burn it to a CD so that you can distribute it as you wish, you may want copyright protection. However, make sure that all the work in the portfolio is yours and that you are not infringing on someone else's copyright. You can copyright the portion of a work that is yours. For work that is not yours, you will need to get the permission of the copyright owner, a topic discussed later in this chapter. If your Web portfolio satisfies these criteria, you may want to post it to your Web site and copyright any original artifacts. You can find specific instructions on how to copyright your work on the U.S. Copyright Office Web site, Circular 66, Copyright Registration for Online Works. You cannot copyright your domain name.

How do I copyright my work?

First, you do not have to register a work with the U.S. Copyright Office to have it copyrighted. Once an original work or artifact is in a fixed, unchanging form, it is copyrighted. In today's digital world, of course, few if any works are ever really fixed or unchanging. Still, an original digital work is copyrightable if it can be perceived as a whole and that whole is distinct from other artifacts of that genre or type. If you decide to register an artifact in your portfolio, you will need to submit a completed registration form, along with a filing fee of $30 and the required copy or copies of the works you wish to register. Online registration is now available. Your work does not have to be published in order to be protected by copyright. Each artifact can be registered separately or as a collection on one application under one title or name. For each application, you must pay $30. You may submit your work as a CD-ROM but not as a floppy disk or zip disk.

How long does copyright protection last?

Copyright protection lasts for the life of the author plus an additional 70 years. For works made for hire (discussed later in this chapter), copyright lasts for 95 years from the date of first publication or for 120 years from the year of creation, whichever comes first. For more information on the duration of copyright, see Circular 15a, Duration of Copyright, on the U.S. Copyright Office Web site. You do not have to renew your copyright for works created on or after January 1, 1978. To let your audience know that your work is copyright protected, you should mark it with the copyright symbol © or the word "copyright," plus your name and the date of first publication (e.g., © 2005 Susan Smith). While marking your work with the copyright mark is optional, it is usually a good idea to do so.

As noted earlier, you don't have to register your work with the U.S. Copyright Office to have it protected; however, doing so provides you with certain benefits, which are listed in Figure 6.3.

Your work is on record, and formally protected from potential copyright infringement.

You have the right to file suit on an infringement claim.

You can collect statutory damages for any infringement claim that is proven.

You can recover attorney fees for any successful infringement claim.

FIGURE 6.3
Benefits of Registering Your Work with the Copyright Office
Source: Herrington (2003), 94.

Can I apply for an international copyright?

There is no such thing as an international copyright that will automatically protect your work throughout the world. Protection from unauthorized use in a particular country depends largely on the laws of that country. If you desire to have copyright protection in a specific country, you should research the protection provided to foreign authors by the copyright laws of that country. It is best to do this research before the work is published anywhere, since copyright protection in that country may depend on when the work was first published. Most countries offer some protection to foreign works under certain guidelines, and these guidelines have been clarified by international copyright treaties and copyright conventions. If those countries have agreed to the provisions of one of the international copyright conventions, then your work will be protected according to the agreements of that convention. If you want additional information in this area of intellectual property rights, including a list of the countries that have copyright agreements with the United States, you can visit the U.S. Copyright Office Web site (www.copyright.gov), International Copyright Relations of the United States, Circulars 38a and 38b.

What protection does copyrighting my work give me?

As noted earlier, the copyright law protects your work from misuse by others. It also provides you, the author of the work, with exclusive rights to any of the activities identified in Figure 6.4.

THE FAIR USE DOCTRINE AND YOUR PORTFOLIO

The part of the 1976 Copyright Act that will play a major role in helping you determine what materials to include in your portfolios and how to use those materials is often referred to as the "fair use doctrine." This doctrine, defined in Section 107 of the act, is complex and has far-reaching implications. What is true today regarding fair use may not be true tomorrow, so it is important to stay current on existing copyright laws. While the language used in defining fair use is not easy to understand, the doctrine attempts to strike a balance between an author's right to own and profit from a work (addressed in points 1 and 4) and the public's right to have the work for educational purposes. The doctrine itself is relatively short and appears in Figure 6.5. To determine whether you are using copyrighted material under the protection of the fair use doctrine, you need to apply the four factors listed in this figure to each use of copyrighted material. Remember, when in doubt, it is always best to get permission in writing to use copyrighted material.

In preparing your portfolios, be aware of the guidelines governing fair use. The three types of fair use that will have the largest impact on your portfolios are the following:

* Educational fair use
* Personal fair use
* Creative fair use

(Permission to use the three types of fair use as stated is granted by the USG Office of Legal Affairs. Copyright and Fair Use <http://www.usg.edu/legal/copyright>)

To reproduce the copyrighted work

To prepare derivative works based on the copyrighted work

To distribute or sell copies or phonorecords of the copyrighted work to the public or to rent, lease, or transfer ownership of this work

To perform the copyrighted work publicly or to display it publicly

---

FIGURE 6.4
Author Privileges Provided by the Copyright Law
Source: U.S. Copyright Office, *Copyright Basics* (Circular 1), p. 1.

"Notwithstanding the provision of sections 106 and 106A, the fair use of a copyrighted work, including such use by reproduction in copies or phonorecords or by any means specified by that section, for purposes such as criticism, comment, news reporting, teaching (including multiple copies for classroom use), scholarship, or research, is **not** (emphasis added) an infringement of copyright. In determining whether the use made of a work in any particular case is fair use the factors to be considered shall include—

1. The purpose and character of the use, including whether such use is of a commercial nature or is for nonprofit educational purposes;

2. The nature of the copyrighted work;

3. The amount and substantiality of the portion used in relation to the copyrighted work as a whole; and

4. The effect of the use upon the potential market for or value of the copyrighted work."

The fact that a work is unpublished shall not itself bar a finding of fair use if such finding is made upon consideration of all the above factors.

FIGURE 6.5
The Fair Use Doctrine
Source: Copyright Law of the United States, Section 107—Limitations on Exclusive Rights: Fair Use, pp. 18–19.

Educational fair use

The fair use doctrine of the 1976 Copyright Act makes it clear that a limited amount of copyrighted material may be used in teaching, scholarship, and research without requiring written permission and without violating copyright law as long as this is done for a nonprofit organization. Educational fair use permits teachers to make multiple copies for classroom use. It can also apply to classroom presentations of your portfolio as a work in progress. For example, if your teacher requires you to present your portfolio design in class, you may want to bring in photocopies of pages from published portfolios for critical review in order to present your own portfolio design ideas. Doing so would be permitted according to educational fair use. Educational fair use also allows Nanette to include copyrighted digital images in her portfolio as long as it is a one-time use, she acknowledges her sources, and she uses these images solely for educational purposes. However, educational fair use does not cover commercial copiers. A major commercial copier once made course packs for classroom instruction at the request of teachers. Such use of copyrighted materials was ruled by a federal district court to be an infringement of copyright.

Over the years, there have been many interpretations of the fair use provision of the copyright law. As noted earlier, the doctrine is complex, and each situation is different, depending upon such variables as the use of the material, the length of the passage quoted, paraphrased, or summarized, and the impact of this use on the market value of the copyrighted work. How much of someone else's work you can use without permission is not an easy question to answer, for each case is different. Your English teacher may have told you that you can use up to 250 words in total (or no more than 5 percent of the original work) from a copyrighted work of prose without getting permission, and this guideline is often used by publishers. However, if the copyrighted work is a song lyric or a poem, as little as one line or two notes may be an infringement of copyright. When in doubt about whether or not you may be in violation of copyright law, get permission from the copyright owner.

Personal fair use

Personal fair use involves using a copyrighted work for learning or entertainment. For example, you may decide to copy two articles from a trade journal in your library and to download two articles from the Internet as part of your research on different portfolio designs. These activities are protected by personal fair use. You can also transfer songs from your personal CD music library (if these CDs have been purchased legally) onto a single CD for your own enjoyment. Doing so is sometimes referred to as "format sharing," which is protected by personal fair use. You cannot, however, make the CD of your personal favorites available to your friends on an Internet site. Doing so would no longer be covered by personal fair use. The Digital Millennium Copyright Act of 1998 (discussed later in this chapter) may change how personal fair use is legally interpreted.

Creative fair use

Creative fair use is the oldest form of fair use, going back at least to the nineteenth century. The principle here is the recognition that knowledge is seamless and that new works are often built on work that has gone before. Often this work is copyrighted, and creative fair use permits an author to use copyrighted work to create new work. For example, the U.S. Supreme Court ruled in 1994 that a rap group's song, "Big Hairy Woman," a parody of the famous Roy Orbison song "Oh, Pretty Woman," was fair use of that work because the rap song resulted in a new creative work, even though that new work was based heavily on Orbison's song (Crawford and Murray 2002, 70). Creative fair use supports the basic goal of advancing learning by encouraging creators to develop new intellectual property. The distinctions between educational fair use, personal fair use, and creative fair use are often blurry, however, and contribute to the complexities of the fair use doctrine.

Photographs, digital images, and copyright

As mentioned earlier, Section 102 of the 1976 Copyright Act lists "pictorial, graphic, and sculptured works" as one of the categories of authorship protected by copyright. Unless you are using such works for academic/ educational use, you will need to get permission to use them. However, if your portfolio is an academic project, you may use digital images/artwork, photographs, and print graphics in it without getting permission as long as you credit each source. Such academic use is protected under the fair use doctrine. If you decide to Web host your portfolio containing copyrighted work, you are no longer protected by the fair use doctrine governing academic use. Your Web site is now available to anyone on the Internet and is not restricted to the academic community. To abide by educational fair use, you should limit access to your Web site and your Web-hosted portfolio by using technology, such as password access, so that your portfolio isn't available to everyone on the Internet. Since the U.S. Copyright Law is amended frequently, you may want to double check the U.S. Copyright website (http://www.copyright.gov) before posting anything on the Internet.

Getting permission to use copyrighted material

If you decide that your use of copyrighted material falls outside of the fair use doctrine, you must get permission to use it. If you want permission to use a Web document, you will often find an e-mail address that you can use for this purpose. If you want to use copyrighted material from a book or journal, you can visit the publisher's Web site, where you will find information on how to request permission to use it. You can also check the copyright notice for the author's name. If you want to use material from a copyrighted work that isn't registered with the Copyright Office, you may find the author's name by contacting the Authors Registry, which maintains a database of authors who have been paid for their work (Crawford and Murray 2002, 74). If you find only the author's name and no other contact information, you may need to conduct an Internet search to get an e-mail address or Web site that you can use to get permission. An e-mail or fax, of course, is the quickest way to get permission. Most authors will want to know how you plan to use their work, so you may end up writing an explanatory letter. Your letter should also include a signature section that the author can sign granting you written permission to use the copyrighted material. Figure 6.6 provides guidelines for writing a permission letter. Considerable information is available in print and on the Internet on how to get permission to use copyrighted material.

1. Address your letter to the copyright holder, whether it be an individual, company (get the name of the correct person), or publisher.

2. Clearly describe the subject matter of the portfolio piece that will include the copyrighted material.

3. Provide the following information about the copyrighted materials you wish to use:
    *   Author(s)
    *   Title
    *   Copyright date
    *   Copy of material for which permission is sought

4. Clearly describe how this information is to be used, including how the information will be reproduced.

5. Include an authorization form with the following information:
    *   Author and title of work
    *   Use
    *   Conditions, if any
    *   Signature line preceded by the words "Permission granted by:"

6. Include a self-addressed stamped envelope if you want a paper copy rather than an electronic response.

FIGURE 6.6
Guidelines for Writing a Permission Letter
Source: Adapted from Carol Simpson, *Copyright for Schools: A Practical Guide*, 4th ed., Worthington, OH: Linworth, 2005.

WORK-FOR-HIRE DOCTRINE

One of the areas of intellectual property that is likely to affect you now and later on in your career is the work-for-hire doctrine. This section examines the work-for-hire doctrine from two vantage points: how it may affect you and your portfolio and how it may affect your work throughout your career.

Work-for-hire and portfolios

If you are hired by a company as either a paid or unpaid intern and you want to use any of the work you created for or contributed to in your portfolio, you should get written permission from the company first. Even though you are using that work for educational purposes, by law you will probably be considered a company employee, even though you may work for only a few hours a week and for only a proscribed period of time. The company owns the copyright to the work that you created unless (and rarely, if ever, does this apply to internship agreements) there is a written agreement between you and the company that gives you copyright privileges. If you work as a part-time or full-time employee and want to use some of your work for your portfolio, the same rule applies. The company, not you, owns the copyright to the work, and you need permission to use any part of it in your portfolio.

Work-for-hire and professional communicators

The work-for-hire doctrine is a complex area of intellectual property rights. Section 101 of the 1976 Copyright Act defines works made for hire as presented in Figure 6.7. We have edited provision 2 of the work-for-hire doctrine to include only those examples that are likely to be created by professional communicators.

Provision 2 also defines the terms "supplementary work" and "instructional text" in some detail. Supplementary works are those creations that support the work of another author. These include graphics, indexes, appendices, and bibliographies. Instructional texts are written documents or graphics prepared for publication to be used in instruction or training.

A work made for hire is—

1. a work prepared by an employee within the scope of his or her employment; or

2. a work specifically ordered or commissioned for use as a contribution to a collective work, as a part of a motion picture or other audiovisual work, as a sound recording, as a translation, as a supplementary work, as a compilation, as an instructional text, as a test, as answer material for a test, or as an atlas, if the parties expressly agree in a written instrument signed by them that the work shall be considered a work made for hire. . .

FIGURE 6.7
The Work-for-Hire Doctrine
Source: U.S. Copyright Office, *Works Made for Hire Under the 1976 Copyright Act* (Circular 9).

Interpretations of the work-for-hire doctrine are made on a case-by-case basis. Whether or not a work is made for hire depends on the relationship between the parties involved. As Herrington notes, interpretations of work-for-hire status are rooted in agency-partnership law (Herrington 1999, 126). For an excellent detailed discussion of the work-for-hire doctrine and agency law, see Herrington, *A Legal Primer for the Digital Age.* Here we cover a few of the broad issues related to the work-for-hire doctrine so that you, as a professional communicator, have a basic understanding of how this doctrine affects the intellectual property that you create. For a work-for-hire relationship to occur, three factors must be in place:

1. Employer control over the work (e.g., the work is done at the employer's place of business, using the employer's equipment)
2. Employer control over the employee (e.g., the employer assigns and schedules employee projects and pays the employee on a regular basis, not a one-time fee)
3. Status and conduct of employer (e.g., the employer provides benefits and withholds taxes.)

Changes in technology are also major factors affecting the work-for-hire relationship. Many professional communicators (and you may be one of them) telecommute part-time or even, perhaps, full-time from a home office or an Internet café. As a result, work-for-hire distinctions regarding place of work are blurry at best, and these distinctions still need legal clarification. It is best for you and your employer to spell out in writing exactly what the work-for-hire agreement covers.

In discussing the work-for-hire doctrine as it relates to employee status, Herrington (1999) identifies 13 elements for defining the employer-employee relationship under agency law. These elements may help you, as a professional communicator, clarify whether your relationship with your employer is likely to be interpreted as that of an employee or an independent contractor under agency law. Herrington also notes that these elements should be viewed collectively rather than individually when you try to determine your work-for-hire status. Figure 6.8 lists these 13 elements. For a full discussion of each of these points, see Herrington (1999).

1. If the hiring party had a right to control the manner and means for creating the product, the likelihood is increased that the work is for hire.
2. The higher the level of skill required, the greater the likelihood that the creator is an independent contractor.
3. Where a hiring party provides instruments and tools to create the intellectual product, the court will find support for determination of a work-for-hire.
4. If the hired party worked at the hiring party's place of business rather than his or her own, this element will lead to a more likely finding of work-for-hire.
5. The longer the duration of the relationship between the two parties, the greater the possibility of a work-for-hire.

Fig. 6.8 continued ....

6. When the hiring party assigns additional projects to the hired party, he or she has more control and is more likely working in the status of employer for purposes of work-for-hire.

7. The more discretion the hired party has over when and how long to work, the more likely he or she is an independent contractor and can maintain control over the work.

8. Hired parties who are paid by the hour, week, or month, rather than by the job, are likely to be employees.

9. When a hired party has a role in hiring and paying assistants, he or she may be legally determined to be an independent contractor.

10. If the work created was something usually within the realm of the hiring party's business, this element could help make a showing that the work was not for hire.

11. When a hiring party does not have an on-going business, it is harder for him or her to claim to be an employer.

12. Hiring parties who pay benefits to hired parties are more likely than not to be employers.

13. Hired parties who are taxed through the hiring party's business are more likely to be employees.

FIGURE 6.8
13 Elements Affecting the Employer–Employee Relationship
Source: From TyAnna K. Herrington, "Who Owns My Work? The State of Work for Hire for Academics in Technical Communication," *Journal of Business and Technical Communication,* (Vol. 13, No. 2), 1999, pp. 136–137, copyright 1999 by Sage Publications, Inc., reprinted by permission of Sage Publications, Inc.

JOINT AUTHORSHIP

You will probably want to include in your portfolio one or more pieces that show your ability to work on a project team, knowing that collaborative projects are a fact of life for professional communicators. If you decide to do so, you should get permission from the other creators. Let's assume, for example, that you were part of a project team with three other people who created a multimedia presentation promoting your campus career center. You wrote the script, someone else created the HTML source code, another created the graphic, and a different person contributed the audio. In its final form, the whole is greater than the sum of its parts, and the multimedia presentation is a seamless work whose value is shared by the four authors who created it. To avoid future conflict, each coauthor should sign a permission form granting every other author the right to use the multimedia project for an educational purpose (i.e., a portfolio piece). If the project has potential commercial value, the permission form should note that any use other than an educational one must be agreed on later by the four authors in a separate agreement. For jointly authored works, copyright lasts for the life of the last surviving author plus 70 years.

If, for some reason, you don't have written permission to use your collaborative project for an educational purpose, you still may do so. Preface the part of your portfolio containing the project with the names of your coauthors and a clear description of your contribution. Using the multimedia project as an example, you would note that you wrote the script for the presentation.

Deep linking

Deep linking is a relatively new issue in the ongoing intellectual property rights debate. It is the practice of creating a link to a subsidiary or secondary page on another entity's Web site, thus bypassing that Web site's home page. You may be using it in creating your electronic portfolio. For example, a section of your portfolio may be devoted to medical communication. In that section is an article you wrote that reports on the growing number of children suffering from asthma. When you discuss different medications for treating asthma, you decide to create a link to a subsidiary page within a pharmaceutical company's Web site rather than linking to that company's home page. That subsidiary page describes a popular asthma medication mentioned in your article. What you have done is create a deep link to the only page on that site that pertains to your article.

While deep linking is common today, companies are beginning to challenge this practice. At stake is advertising revenue. Because companies sell advertising space on their Web sites, the risk of losing advertising dollars increases when visitors bypass the home page where advertising appears for more content-specific secondary pages. Since Web sites often log the number of visitors who view the advertisement, the number of visitors who navigate the site the way it was designed will have a direct bearing on how much advertisers are charged. Most court decisions to date support the ruling that deep linking is not a violation of intellectual property protection. A prominent case involving deep linking was *Ticketmaster v. Tickets.com*, where Ticketmaster claimed that tickets.com, a competitor, was violating intellectual property laws by deep linking to Ticketmaster's site. A federal judge ruled that there was no copyright infringement since no copying was done (Markel 2002, 78). If you decide to pursue a career as a Web developer, you should stay informed on decisions that affect deep linking.

THE DIGITAL MILLENNIUM COPYRIGHT ACT OF 1998

The Digital Millennium Copyright Act (DMCA) extends the intellectual property rights described in the 1976 Copyright Act to include digital works. The DMCA's purpose is to protect copyright owners in the digital age from digital piracy. It includes the five provisions identified in Figure 6.9. For example, you decide to create a Web-hosted version of your portfolio and protect access to it with password protection technology. Someone bypasses the technology, views your portfolio, and downloads your copyrighted files. Under the DMCA, that activity is illegal because this person has bypassed a technology designed specifically to protect a digital work.

When the DMCA was signed into law in 1998, the Motion Picture Association of America and the Recording Industry Association of America were very happy about the protection they received from digital piracy because the home video market is so profitable. Librarians and academics, to name two groups, however, are not happy because they see the DMCA as eroding the fair use principle and establishing new barriers to academic research. For example, the wording of the DMCA, broadly interpreted, could make it illegal to view or copy Internet material without prior permission regardless of the intent of the user. The DMCA could also limit scholarship that might involve assembling databases from Internet material (Foster, 2003, 2). The provisions of the DMCA are complex, and it is not the intent of this chapter to discuss this act in detail. You can

---

Title I, the "WIPO Copyright and Performance and Phonograms Treaties Implementation Act of 1998," implements the World Intellectual Property Organization treaties.

Title II, the "Online Copyright Infringement Liability Limitation Act," limits the liability of online service providers for copyright violations when engaging in certain types of activities.

Title III, "The Computer Maintenance Competition Assurance Act," permits the copying of a computer program if it is being done to re-activate the computer after maintenance or repair.

Title IV has six copyright provisions that pertain to distance learning, library use (copies) of digital materials, "webcasting" of sound recordings on the Internet, and collective bargaining agreement obligations affecting the transfer of rights in motion pictures.

Title V, the Vessel Hull Protection Act," establishes new protection for the design of vessel hulls.

FIGURE 6.9
The Main Provisions of the 1998 Digital Millennium Act
Source: U.S. Copyright Office, *The Digital Millenium Copyright Act of 1998*, p. 1.

view the entire act by using the search feature on the Copyright Office Web site. The DMCA is likely to be hotly debated for many years, and professional communicators should be familiar with the intellectual property issues covered by the act.

INFORMATION LIABILITY

The term "information liability" (LaPlante 1986, 37), coined in the 1980s, describes how writers can be held accountable for injuries that might be caused by information they have written. Information liability can encompass such works as user documentation, misleading or incorrect graphics, marketing pieces, informational reports (e.g., product descriptions), and safety messages. While you may be creating your portfolio initially as an academic requirement, you will be using it for job interviews and, perhaps, for promotion. As you develop it, make sure that the pieces you include show your understanding of information liability.

The following scenario shows how information liability might work. You may decide to include in your portfolio a section from user documentation that you wrote for a course or on the job. If you decide to include a section from a user manual, check to make sure that your instructions are accurate and clear and, most of all, protect the user from potential harm. If poorly worded instructions or safety messages result in lost or corrupted computer files, that's unfortunate; however, injuries caused by unclear, inaccurate, or incomplete instructions can have legal repercussions for professional communicators, as described in the scenario in Figure 6.10. Publications like the *Product Safety & Liability Recorder* are valuable resources with information on topics that pertain to information liability.

PORTFOLIOS AND ETHICAL ISSUES

So far, Chapter 6 has addressed several legal issues that you might face as you build your portfolio. You may also find that a decision may be sound legally but incorrect ethically. This part of the chapter discusses some of the ethical issues that may influence the choices you make. For example, you may be interviewing for a job and an employer is very impressed with the graphics in a team project included in your portfolio. While your overview or introduction to the project may have clearly stated that your contribution consisted of writing the text, the employer has missed this point when reviewing the document. While you are not legally obligated to tell the employer that you didn't create the graphics, ethically you should say that they were created by another project team member. Building your portfolio provides you with an excellent opportunity to reflect on the values on which you will base your professional career and reputation. In short, your portfolio will help you develop your own professional code of ethics, a code that expresses your personal, religious, cultural, and professional values.

Acknowledging your sources

In discussing intellectual property rights, we noted that you are legally obligated to follow existing copyright laws when borrowing from another source and that written permission is needed to use copyrighted

---

An injured user sues the party who sold him or her the product; the seller sues the company that made the product; the company sues the employer, in this case the writer, who wrote the user documentation.

FIGURE 6.10
Information Liability Scenario
Source: Smith and Shirk (1996), 191.

information if the fair use doctrine does not apply. Even when you don't quote directly from that source, you are ethically obligated to credit it for any information that is paraphrased or summarized unless that information is common knowledge or in the public domain.

Avoiding plagiarism

Ever since grade school, you have probably been told never to plagiarize, or present someone else's published or unpublished ideas or words as your own. As you build your portfolio, make sure that you follow the guidelines your school or university has published on how to avoid plagiarism. Never lift documents from the Internet and present them as your own work. Remember that even unintentionally presenting someone else's work as your own is still plagiarism. In short, plagiarism is a form of copyright infringement.

Professional organizations and ethical codes

As a professional communicator, you may already be a member of one or more professional associations (see Chapter 8 for a list of some of the major associations). Almost all professional associations have published codes of ethics that their members follow. The Code of Ethics for the Society for Technical Communication (STC), for example, is presented in Figure 6.11.

The STC Code of Ethics, like many ethical codes, echoes values that are thousands of years old. Perhaps you read *The Republic of Plato* in a college philosophy course and remember Plato describing a virtuous person as someone having the qualities of justice, reason, wisdom, and temperance or self-control (*The Republic of Plato* 1962, 139). You may also remember Plato's famous allegory of the cave, where the cave serves as a metaphor for human nature. In the cave are prisoners (the unenlightened) and puppeteers (the enlightened). The prisoners are forced to watch shadows and images on the cave walls that are projected there by the puppeteers. In the dimly lit cave, these shadows and images distort the real objects they represent; today we could interpret Plato's cave allegory as an early admonition against deceptive advertising, a topic addressed in the section on "Honesty" in the STC Code of Ethics.

In discussing ethics and technical communication, Herrington identifies three popular approaches for testing ethical behavior: universalization, common practice, and the utilitarian test (Herrington 2003, 7). "Universalization" is a central principle in Immanuel Kant's ethical thought system, for he believed that certain actions were by their nature absolutely right or wrong. Kant (1724–1804) referred to this principle of universalization as the "categorical imperative." While the comments made here are a great oversimplification of his philosophy, for Kant categorical imperatives are universal and always ethical, and they require you to do the right thing (take the right action) (Kant, 1964, 70). In very broad terms, the STC's descriptions of the ethical principles of honesty and quality, in particular, have much in common with Kant's categorical imperative. For example, placing graphics in your portfolio that you didn't create or whose ownership you didn't acknowledge would be a violation of Kant's ethical system.

Very briefly, the "common practice test" for ethical conduct treats activities or behavior as ethical if they are common to the profession. This view is very different from Kant's categorical imperative. For example, it is common practice in advertising to let the consumer know that a physician in a television commercial is not a real physician but an actor playing the part. If your company decides to mislead the consumer into thinking that the actor is a physician in real life, then the company is not following the common practice test for your industry. The principle behind the common practice test can be seen in the section on "Honesty" in Figure 6.11. For example, if you exaggerate or misrepresent the contributions you made to a group project in your portfolio, you would not be following the common practice test of the technical communication profession.

The "utilitarian test" for ethics weighs the good or benefit to be gained from taking an action against the harm or detriment that could result from that action. One proponent of utilitarianism is the English philosopher John Stuart Mill (1806–1873). In "Utilitarianism" (1998), Mill measures the rightness or wrongness of an act by the degree of happiness the act brings to the greatest number of people. Herrington uses the following example to

---

Ethical Principles for Technical Communicators

As technical communicators, we observe the following ethical principles in our professional activities.

Legality

We observe the laws and regulations governing our profession. We meet the terms of contracts we undertake. We ensure that all terms are consistent with laws and regulations locally and globally, as applicable, and with STC ethical principles.

Honesty

We seek to promote the public good in our activities. To the best of our ability, we provide truthful and accurate communications. We also dedicate ourselves to conciseness, clarity, coherence, and creativity, striving to meet the needs of those who use our products and services. We alert our clients and employers when we believe that material is ambiguous. Before using another person's work, we obtain permission. We attribute authorship of and ideas only to those who make an original and substantive contribution. We do not perform work outside our job scope during hours compensated by clients, or employers except with their permission; nor do we use their facilities, equipment or supplies without their approval. When we advertise our services, we do so truthfully.

Confidentiality

We respect the confidentiality of our clients, employers, and professional organizations. We disclose business-sensitive information only with their consent or when legally required to do so. We obtain releases from clients and employers before including any business-sensitive materials in our portfolios or commercial demonstrations or before using such material for another client or employer.

Quality

We endeavor to produce excellence in our communication products. We negotiate realistic agreements with clients and employers on schedules, budgets, and deliverables during planning. Then we strive to fulfill our obligations in a timely, responsible manner.

Fairness

We respect cultural variety and other aspects of diversity in our clients, employers, development teams, and audiences. We serve the business interests of our clients and employers as long as they are consistent with the public good. Whenever possible, we avoid conflicts of interest in fulfilling our professional responsibilities and activities. If we discover a conflict of interest, we disclose it to those concerned and obtain their approval before proceeding.

Professionalism

We evaluate communication products and services constructively and tactfully, and seek definitive assessments of our own professional performance. We advance technical communication through our integrity and excellence in performing each task we undertake. Additionally, we assist other persons in our profession through mentoring, networking, and instruction. We also pursue professional self-improvement, especially through courses and conferences.

FIGURE 6.11
Code of Ethics of the Society for Technical Communication
Source: Used with permission from the Society for Technical Communication, Arlington, VA, U.S.A.

illustrate the utilitarian test. If a technical communicator working for a chemical company printed a brochure with inaccurate information, he or she would be applying the utilitarian test to decide whether or not to mail the brochures by weighing the potential harm that the inaccurate information might cause against the cost of reprinting an error-free version of the brochure (Herrington 2003, 9). For a more detailed discussion of these three ethical tests see Herrington (2003, Chapter 2). In closing, the decisions you make in creating your portfolio provide an excellent opportunity to examine the ethical code upon which you will build your professional reputation.

SUMMARY

As a professional communicator, you will face a range of legal and ethical issues as you design your paper and electronic portfolios. Chapter 6 discusses such key areas as intellectual property rights, work for hire, joint work, agency and intellectual property, information liability, and ethical issues. As you begin building your portfolio, you will discover that most or all of these areas will influence the decisions you make regarding the content, design, and presentation of your portfolios. Knowing this information will help you make wise and legally defensible decisions.

Chapter 6 also encourages you to reexamine your own code of ethics as you move through the portfolio-building process. One way to start is to become familiar with the ethical codes of the professional organizations to which technical and professional communicators belong.

ASSIGNMENTS

Assignment 1: *Checking Your Portfolios for Copyrighted Materials*

Review the graphics and/or digital images in your portfolios, either as stand-alone works or as works that you have brought into a written document, PowerPoint presentation, or multimedia presentation. How many of them are protected by copyright? Select some sample graphics and do a Google search using keywords such as "graphics and public domain" or other phrases, some general and others more specific, that will allow you to locate graphics in the public domain. Evaluate these graphics to see if you can use any of them in place of graphics in your portfolios that are protected by copyright.

Assignment 2: *Evaluating Ethical Codes*

Search the Internet for the Web sites of two professional communication associations and examine their codes of ethics. You might begin your search by reviewing the Web sites of the professional associations listed in Table 8.5 in Chapter 8. Compare the two codes of ethics for the general principles they espouse. Write a short analysis noting how these codes are similar and different.

REFERENCES

Crawford, Tad and Kay Murray. *The Writer's Legal Guide: An Authors Guild Reference*, 3rd ed. New York: Allworth Press, 2002.

Fishman, Stephen. *The Public Domain: How to Find Copyright-Free Writings, Music, Art and More*. Berkeley, CA: Nolo Press, 2000.

Foster, Andrea. "Library Groups Say Sweeping State Copyright Laws Could Stifle Teaching and Research." *Chronicle of Higher Education* 49, no. 32 (April 1, 2003): 1–5

Herrington, TyAnna. "Who Owns My Work? The State of Work for Hire for Academics in Technical Communication." *Journal of Business and Technical Communication* 13, no. 2 (April 1999): 125–153.

Herrington, TyAnna. *A Legal Primer for the Digital Age*. New York: Pearson Longman, 2003.

Kant, Immanuel. *Groundworks of the Metaphysic of Morals*. Translated by H. J. Paton. New York: Harper & Row, 1964.

LaPlante, Alice. "Liability in the Information Age." *InfoWorld* 8, Issue 33 (August 18, 1986): 37–38.

Markel, Mike. "Deep Linking: An Ethical and Legal Analysis." *IEEE Transactions on Professional Communication* 45, no. 2 (2002): 77–83.

Mill, John. S. "Utilitarianism." In *The Utilitarians*. Garden City, NY: Dolphin Books, 1998.

The Republic of Plato. Translated by F.M. Cornford. New York: Oxford University Press, 1962.

Simpson, Carol. *Copyright for Schools: A Practical Guide*, 4th ed. Worthington, OH: Linwood Publishers, 2005.

Smith, Howard and Henrietta Shirk. "The Perils of Defective Documentation: Preparing Business and Technical Communicators to Avoid Products Liability." *Journal of Business and Technical Communication* 10, no. 2 (1996): 187–202.

Society for Technical Communication. *Ethical Principles for Technical Communicators.* Washington, DC: Society for Technical Communication, 2005 http://www.stc.org/policyStatements_ethicalPrinciples.asp (accessed January 24, 2005).

Twain, Mark. *Mark Twain's Notebook*, ed. Albert B. Paine. New York: Cooper Square, 1972, p. 381.

University System of Georgia Office of Legal Affairs. "The Three Kinds of Fair Use: Creative, Personal, and Educational." *Regents Guide to Understanding Copyright and Educational Fair Use.* http://www.usg.edu/legal/copyright (accessed April 7, 2005).

U.S. Copyright Office. *The Digital Millennium Copyright Act of 1998.* Washington, DC: Library of Congress Copyright Office, 1998. http://www.copyright.gov/legislation/dmca.pdf (accessed April 6, 2005).

U.S. Copyright Office. *Copyright Basics* (Circular 1). Washington, DC: Library of Congress Copyright Office, 2003. http://www.copyright.gov/circs/circl.html (accessed August 13, 2004).

U.S. Copyright Office. *International Copyright Relations of the United States* (Circular 38a). Washington, DC: Library of Congress Copyright Office, 2003. http://www.copyright.gov/circs

U.S. Copyright Office. *Highlights of Copyright Amendments Contained in the Uruguay Round Agreements Act (URAA)* (Circular 38b).Washington, DC: Library of Congress Copyright Office, 2003. http://www.copyright.gov/circs

U.S. Copyright Office. *Copyright Law of the United States. Section 107—Limitations on Exclusive Rights: Fair Use* (Circular 92). Washington, DC: Library of Congress Copyright Office, 2004.

U.S. Copyright Office. *Works Made for Hire Under the 1976 Copyright Act* (Circular 9). Washington, DC: Library of Congress Copyright Office, 2003. http://www.copyright.gov/circs

APPENDIX A: SUMMARY INFORMATION SHEET ON LEGAL AND ETHICAL ISSUES COVERED IN THIS CHAPTER

1. What main categories of intellectual property are protected by the 1976 Copyright Act?

    Answer: The 1976 Copyright Act protects the following major categories of original authorship:

    * Literary works
    * Musical works
    * Dramatic works
    * Graphics
    * Motion pictures
    * Sound recordings
    * Architectural works

2. What works are not protected by the 1976 Copyright Act?

    Answer: This act does not protect the following categories of work:

    * Works that are not in a tangible, fixed form
    * Works that consist entirely of common property
    * Works in the public domain

3. What is public domain?

    Answer: The term "public domain" refers to works that are no longer protected by copyright laws or works that do not meet copyright requirements. All work produced by the U.S. government is considered in the public domain.

4. Must you register your intellectual property with the U.S. Copyright Office to have it protected by copyright laws?

    Answer: No, but registering your work does provide certain benefits, such as defending yourself more easily from copyright infringement.

5. Should I register my portfolio(s) with the U.S. Copyright Office?

   Answer: Probably not, since your portfolios are evolving documents and are not a fixed form of expression. You may want to copyright protect individual artifacts within your portfolio.

6. What is fair use, and what are its limitations?

   Answer: The fair use doctrine of the Copyright Act of 1976 is one of the more important and one of the more complex areas of intellectual property law. Each use is different and is governed in part by the type of copyrighted work, the purpose or use of that work, the amount borrowed, and the impact that the use has on the market value of the original. Using a limited amount of copyrighted material without written permission for educational purposes is normally considered fair use. For example, you may use a limited amount of copyrighted material in your portfolio if your portfolio is being developed for a class assignment. In all other cases, you will need written permission to use this material. How extensively the Digital Millennium Copyright Act of 1998 will affect fair use remains to be determined.

7. What conditions must be present for a work-for-hire relationship to exist?

   Answer: Like the fair use doctrine, the work-for-hire doctrine is complex. In a work-for hire relationship, the U.S. Copyright Office has stated that three conditions must exist:

   * Employer control over the work
   * Employer control over the employee
   * Status and conduct of the employer (i.e., provides employee benefits and/or withholds employee taxes)

8. Can I include work in my portfolio that is jointly authored? What safeguards should I use?

   Answer: Yes. To avoid future conflicts regarding authorship, you should get written permission from the other authors. That permission should clearly identify the authors, describe the contribution of each, and specify the conditions under which the portfolio piece can be used. All authors should sign the permission agreement. If this is done, then each author can use the jointly created work in his or her own portfolio.

9. Is there such a thing as international copyright law for intellectual property?

   Answer: No. Copyright laws differ from country to country, although most countries offer some protection to foreign works under guidelines established by international copyright treaties and copyright conventions (e.g., the Berne Convention). For more information, consult the International Copyright Relations of the United States (Circulars 38a and 38b), found on the U.S. Copyright Office Web site (www.copyright.gov).

10. What publications will help me keep current on product liability issues?

    Answer: There are many publications that you can consult. A few of the major publications are listed below and are found in most libraries.

    * *Product Safety and Liability Reporter*
    * *United States Code*
    * *Product Liability Desk Reference*

11. What are some of the ethical issues that professional communicators should be aware of?

    Answer: As a professional communicator, you should first become familiar with the ethics codes of the professional organizations that will directly affect the type of work you do. You should develop your own ethical code as a professional communicator that is consistent with your profession's code and that addresses such principles as honesty, fairness, and being a user advocate.

12. What is the Digital Millennium Copyright Act of 1998, and how will it affect what I include in my portfolios?

    Answer: The Digital Millennium Copyright Act (DMCA) is likely to continue to have a major impact on intellectual property issues. From deep linking to fair use, the DMCA is changing the landscape of intellectual property and how researchers use that property. In brief, the DMCA protects copyright owners of digital works from digital piracy and makes it illegal for anyone to bypass technologies created to protect copyrighted intellectual property that is digital in nature.

7 Getting Feedback: Responding to and Revising Portfolios

When the portfolio process first begins, there are all types of ideas that float around. We have grandiose ideas about what our portfolios should be like, including all the bells and whistles. However, in discussing our work with our peers, we get a more grounded center and we can more easily draw our attention to what is most important. Judith

INTRODUCTION

This chapter gives advice on finishing the first draft of your paper and electronic portfolios, providing guidelines for an effective peer response from classmates and/or colleagues. It also discusses feedback for use in revision and the ways you might reshape and polish your portfolios for professional presentation. Chapter 7 covers the following topics:

* Importance of feedback
* Feedback during invention
* Feedback on developed drafts
* Developing criteria for paper and electronic portfolios
* Establishing feedback criteria for portfolio workshops
* Conducting a workshop session
* Processing and implementing feedback
* Editing

IMPORTANCE OF FEEDBACK

In the chapter-opening quotation, Judith discusses the importance of getting feedback from others. The creation of your portfolios is often a messy process of revision that involves many changes. Judith describes the way she was able to shape her ideas with the help of an audience of interested readers willing to give her productive feedback. As we have emphasized throughout this book, your portfolios do not exist in a vacuum. You can gain valuable insights for recursive revision when you bring others into the processes of composing from invention through final draft. We incorporate feedback into our students' projects throughout the portfolio-building process. We recommend early feedback on your proposals and initial ideas, feedback from mentors in the field, and feedback on finished drafts before producing your final products. Our students had structured workshop/feedback sessions as part of their class. However, if you are not in a classroom setting or want feedback beyond the classroom (something we strongly recommend), you can still elicit a response from others. Colleagues, potential employers, professional mentors, and general users can all provide valuable perspectives throughout the development and revision of your portfolios. Bringing in outside readers is the only way to know how your audience will react to and interpret your portfolios. This type of productive response will help you revise your portfolio more effectively.

EXERCISE 7.1 IDENTIFYING AND PLANNING FOR FEEDBACK

Identify the ways you plan to get feedback on your portfolios.

1. Identify ways you might participate in peer response groups with other students or colleagues who are also working on their portfolios.
2. Identify a professional mentor (a colleague or someone from your potential professional field) who would be willing to give you feedback throughout the process.
3. Set up a timeline for feedback sessions or mentor correspondence during the different processes of composing and designing your portfolios (early, middle, and final versions).

FEEDBACK DURING INVENTION

Throughout this book, we have described many ways to generate ideas for your portfolios. Feedback during the process of invention can help you to shape and refine your ideas and resolve the tension between the fantastic and the realistic. Some students find early feedback helpful because they are stuck or unable to find common threads in their work. Many students, like Judith, start with "grandiose ideas" that need to be grounded in reality. Judith describes the way early feedback helped her to modify her ideas to create something that worked:

> Showing our ideas in these beginning stages gives us a better idea of our audience, which is desperately needed in the document conversion process as well as in the creation of our portfolios.

She goes on to explain the transformation of her original metaphor into an image and a theme that became the driving force for her portfolios:

> In my initial stages I was trying to work on a metaphor for my portfolios. I had in my mind an angel-type figure flowing over a scroll creating words with her breath. Now I thought it was an absolutely terrific idea. However, when presenting it to others, they were not half as excited as I was about it—so there went the idea. However, during my presentation, my peers did like the idea of using a silhouette, and people commented on the concept of defining colors and what they represented to me. I ended up using the colors as my metaphor. With all of this input, my idea was born.

Judith's original angel figure came across as a bit too whimsical. She thought that it gave her portfolio a theme of "ethereality and poetry," but her audience did not feel that it presented a serious, professional image. Her final silhouette image, presented in Figure 7.1, shows her attention to detail and design along with her technical skills. This silhouette, originally only a minor feature in Judith's portfolio, became a major one that eventually appeared on her home page and in her introduction.

Judith originally presented the colors to show connections to her personality, but through discussion her peers helped her realize that the colors themselves might act as a guiding thread. During her proposal presentation, she created a collage of images and textures representing the range of colors for each section

FIGURE 7.1
Judith's Final Silhouette Image
Source: Used with permission from Judith Dickerson.

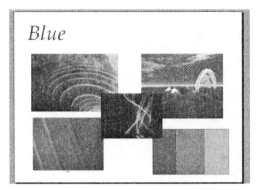

FIGURE 7.2
Judith's Original Color Collage
Source: Used with permission from
Judith Dickerson.

(see Figure 7.2). Although she did not initially intend the colors to come to the forefront, they ended up defining the portfolios and dividing her sections. At the time Judith presented this color collage, she was brainstorming about design features. However, she made such distinct connections between the colors, character traits, and areas of specialization that the collage was strong enough to act as a grounding metaphor for her portfolios. She pulled the whole thing together by incorporating the colors into the silhouette image to form a comprehensive representation. It was through feedback from her peers that she was able to discover this connection.

While Judith ended up radically changing her ideas as a result of audience feedback, Joy used feedback during invention to modify and simplify an idea that was originally very complicated. Early on, Joy presented the idea of a kaleidoscope as a dominant image. Originally, she presented multifaceted images that were somewhat confusing and visually too busy (see Figure 7.3) and proposed changing the color schemes for each section to correspond with particular qualities. Her peers encouraged her to simplify and push for a stronger connection between the image and her work. She says:

> I learned that using the image of the kaleidoscope would have been too busy as a graphic. But the concept of triangulation and multifaceted perspective (as in the three mirrors inside a kaleidoscope) was encouraged and developed as a result. Also, I was encouraged to explain to my audience why I used the image to help them better understand the connection.

Joy modified her original idea into a sleek, simple design that functioned as a logo throughout her work (see Figure 7.4). Her home page and introduction eventually included a short explanation of the connection between the kaleidoscope image and the triangulation of her skills. Joy modified her original design once her

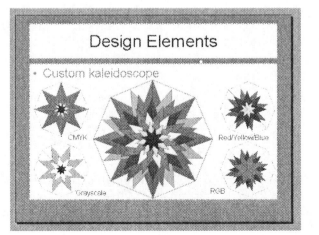

FIGURE 7.3
Joy's Original Design
Source: Used with
permission from Joy
Leake.

FIGURE 7.4
Joy's Final Design
Source: Used with permission from Joy Leake.

audience helped her realize that the changing colors and complexity might distract readers. Her new design created a clean image and a way to easily brand herself through her logo and slogan, which are found throughout her portfolios, on her business cards, and on her CD cover.

Angela, on the other hand, came to her original feedback session with several possible ideas but unsure about which direction to take. She spent quite a bit of time defining her professional identity and looking for images to reflect these qualities. She says:

> My theme originated in our first class feedback session. My original idea was to depict sincerity and truth through various Southern visuals. I wanted to relate my quest for truth and community understanding to the unpretentious honesty of Southern culture.

She presented images of rocking chairs on porches, beautiful oak tree canopies draped in Spanish moss, and iron skillets. Although these images embodied Southern culture, her audience did not necessarily perceive the personal qualities she hoped to convey. Her peers indicated that the "front porch rocking chairs in no way resembled her personality." Instead, they saw her as honest, forceful, and professional. To them, the rocking chairs and the Southern theme reflected a laid-back style that did not reflect her strong work ethic. During the feedback session, her peers collaboratively brainstormed with her to come up with images and themes that seemed more fitting. Her final image—light—emphasized her connection between technical communicators as "bringers of light" and clarity along with a correlation between light and truth. As Angela admits, her ideas came out of a collaborative process as she tried out her ideas on others. As she says, "My willingness to accept their opinions led to a metaphor that suited my conception perfectly."

Other students, such as Nanette, worked closely with a professional mentor while creating the portfolios. Some of his feedback prompted her to make important decisions regarding the length and number of documents. She also realized that she needed an objective outsider to help her come up with a new idea rather than struggling to modify an old one. Sometimes, as writers and designers, we get so close to our own work that we doggedly hold on to ideas merely because we have created them. Nanette describes how her mentor helped her to move beyond this circular struggle:

> I met with my mentor that week, and he also (along with classmates) agreed that my portfolio needed more "spark" to it. My change occurred when I showed him my page that I had designed with the colorful street signs and arrows. He said, "Use this as your theme," and I said that I was hesitant to make changes. He said that I had a choice: either I can spend ten more hours trying to make what I have work or use the same ten hours to create something new.

Nanette convinced herself early on that she wanted to use an existing document to create her home page and theme. She originally thought this would be a good starting place since she had put so much time and energy into the original document. She was so determined to make it work that she lost sight of the fact that the document did not even reflect her current professional identity or skills. It took the feedback from her professional mentor to get her to "go back to the drawing board" and let go of her earlier ideas.

Peer feedback is valuable for larger thematic issues, but it also helps to resolve issues of selection and categorization. For example, when Joy presented an early version of her table of contents, her marketing material was spread throughout the portfolio in a genre arrangement. In her feedback session, someone suggested that she group all of these pieces together to demonstrate a "marketing campaign" since she had several different kinds of documents (Web pages, brochures, business cards, press releases) for the same nonprofit organization. By grouping them together rather than in discrete genre sections, she was able to present her strong marketing persona as well as demonstrate her diversity and skills. This feedback encouraged her to change the overall structure of her portfolio.

We also found that during these early feedback sessions, students helped one another to inventory and expand their technological skills. Often, viewing samples from others demonstrated new ways to approach document design and presentation. For example, Tom learned valuable lessons about creating documents in a new format for faster access. Wilda shared information about working with templates, and Sarah demonstrated how a PowerPoint presentation could be designed to imitate an active Web site. The process of creating your portfolios might engage you in learning new software applications or finding new ways to present material. Discussion with knowledgeable peers during the invention phase can often help you see a path that you might not perceive otherwise.

FEEDBACK ON DEVELOPED DRAFTS

Although revision takes place throughout all phases of creation, this section focuses on revising the nearly finished drafts of your paper and electronic portfolios. This takes place after you have revised your individual documents for portfolio quality (see Chapter 4). Once you have decided on your contents and themes, you should begin to develop mature drafts of your portfolios for feedback. This means pasting up sample pages of the paper portfolio and creating a working template of the electronic version. At this point, you should also have a clear idea of the table of contents, as well as finished drafts of your section introductions and transitions. You will need to find a forum and a way of presenting the portfolios so that others can respond collectively to your developing drafts. As we said earlier, if you do not have access to a classroom setting, you will have to find ways to set up your own peer response sessions (use the contacts and timeline you generated in Exercise 7.1 to structure your sessions). It may be necessary to modify the questions or format based on your particular situation and audience. Our students presented their work in 15-minute response sessions in which they displayed the paper portfolios and projected the electronic versions. These sessions were followed by 15–20 minutes of verbal and written feedback focusing on particular criteria. At this point in the process, respondents are encouraged to point out both strengths and areas that need development. These might include issues of writing, style, design, or navigation.

We find it most productive to set up final criteria that guide respondents through this process. It is important to recognize the purpose of response and feedback in this setting. It is not for evaluation—as in a purely academic setting between a teacher and a student—but should focus on helping students move toward productive revision of their work. We advocate a balance between pointing out strengths and areas that need development. We also feel that the most effective comments include both identification and explanation. It is not enough to say, "This is interesting" or "Move this section." Effective evaluation involves identifying specific places in the text and giving reasons along with ideas for improvement. In order to respond effectively, you should practice this kind of productive feedback.

Our model relies on the terminology used by response theorists Richard Straub and Ronald Lunsford in their study on response styles and criteria. They state that there are two perspectives from which to analyze students' comments (although they use this model to analyze teachers' comments on students' papers, we use it to train students in effective responding strategies): the focus and the mode. As they say, the **focus** "identifies *what* a comment refers to in the writing," and the **mode** "allows us to examine *how* the comment is framed." Focus comments concentrate on issues such as ideas, development, wording, organization, and style. The mode of the comment "characterizes the image a teacher creates for herself and the degree of control she exerts, through that comment over the student's writing" (Straub and Lunsford 1995, 158). Their study, along with others in the field, suggests that the most effective modes encourage writers to re-see their ideas and texts rather than commanding or directing them to change discrete units. Comments delivered in the form (mode) of "questions," "reflective statements," and "suggestions" are more productive than those delivered as "corrections," "evaluations," and "imperatives." We encourage you to think carefully about your role when responding to your classmates' or colleagues' portfolios, along with the way you phrase your comments. Remember, it is not the responder's role to change or appropriate the work of others but to help them find the best way to communicate their ideas and demonstrate their skills. When responding to the texts of others, consider yourself an interested, knowledgeable reader who is reporting your experience while reading their work.

DEVELOPING CRITERIA FOR PAPER AND ELECTRONIC PORTFOLIOS

In order to define a strategy for response, it is important to establish criteria prior to your workshop session. Undoubtedly you have come across elements of quality while designing your portfolios. By the time you are ready to present them in a workshop or distribute them for feedback, you should have a strong sense of what your community values in terms of content, design, and usability. Although there is some crossover between the paper and electronic portfolios, we provide a starting point for developing your criteria. These versions may not necessarily be identical, but both portfolios should have a consistent design and theme so that they work as a package.

Paper portfolios

To review, the paper portfolio is targeted to a live audience. You will generally use it during an interview or in a setting where you are present to explain it. Occasionally, readers will flip through it on their own, but it is used primarily to create talking points and should be flexible to be tailored to different situations. Generally, our students use an oversized zippered case (with a handle for easy carrying). However, one of our students, Wylie, modified this model and created his paper portfolio to imitate a magazine in size, layout, and presentation. Whatever vehicle you choose, there are some common elements that organize your work like a book with a table of contents, section divisions, introductions, and context statements for sections and individual documents.

Our students also provided the following tips when constructing their paper portfolios:

* Use plastic page covers to protect your work as it is handled.
* Print all final documents on high-quality, acid-free paper.
* Use permanent, acid-free, archive-quality glue-stick, spray adhesive, or double-sided tape (rather than liquid glue).
* Choose a portfolio binder (but not one that is too heavy or cumbersome) that is larger than your documents so that you will have room for design elements and context statements.
* Choose a portfolio binder with the handle on the zipper side rather than the binder side. If the handle is on the binder side, the weight of the work combined with the pressure of the handle on this side loosens the fasteners over time.
* Back your documents with layers of colored paper to give them visual depth and to pull in your theme and design. You can also use colored paper to delineate different sections.
* Consider placing tabs on each section for presentation flexibility and organization.

* Use a paper cutter or utility blade rather than scissors to cut background paper and design elements.
* Document context statements can be printed on business card–size paper.
* Find a way to represent longer documents (such as manuals or reports) in brief in the portfolio and bring a full copy along for the interview or scan it into your electronic version. You might choose the table of contents and a couple of representative pages that demonstrate your skills.
* Include copies of your resume, business cards, and a CD copy of your electronic portfolio.

Electronic portfolios

In many ways, your electronic portfolio will be similar to your paper portfolio (refer to Chapter 5 for a detailed discussion of electronic portfolios). For example, you will still have a table of contents (links), an introduction (on your home page), and transitional sections. However, the online format creates some differences. First and foremost, your electronic portfolio will be dynamic and extremely flexible. You have the opportunity to work with layers of information and guide your readers/users in ways that the paper portfolio does not provide. Although you might have the chance to present your electronic portfolio live, generally it will be viewed when you are not present. Many job seekers use their electronic portfolio as a supplement to the paper portfolio or offer it as a leave-behind item that provides a more specific portrait for a closer look. This form gives potential employers time to actually read their writing and carefully consider their particular skills. Because of the portability of the electronic portfolio, you can also include longer documents, interactive elements, and live action (such as audio and video components) in this format. You will need to consider issues such as navigation and technological access, along with the general issues discussed in the section on the paper portfolio. It is important to conduct user studies to gather feedback on how others interact with your electronic document. Carol Barnum (2002, 7) generally defines usability as "the art and science of making documents and products usable, useful, and most of all satisfying." This includes the user's perception of

* Friendliness
* Ease of use
* Usefulness
* Ease of learning
* Ease of relearning

Although usability testing can involve complex, long-term research and data analysis, you can certainly apply these criteria to help define how potential users might interact with your electronic portfolio. Like any form of response, usability testing involves the "interaction of a person with a product" (Barnum 2002, 7) and exposure of your work to a live audience.

Here are some other considerations when designing your electronic portfolio:

* Create a clear home page with a simple, intuitive navigation system.
* Add thumbnails of documents and graphics for another dimension of visual appeal and to provide an additional link to full documents.
* Include technical instructions for your user, including issues such as the best viewing size and the software programs needed.
* Encourage others to try out your electronic portfolio and run it through a modified usability test (for more information, see Barnum's [2002] detailed discussion of usability testing).
* Try out your electronic portfolio on several computers. Often you will find that elements such as screen size, software, or spam-blocker programs will affect the way the portfolio is viewed.
* Establish consistency between your paper and electronic portfolios in terms of color, design, and theme.
* Include "back" buttons or static frames to make sure that your audience does not get lost within the electronic portfolio.

ESTABLISHING FEEDBACK CRITERIA FOR PORTFOLIO WORKSHOPS

It is important for you and your audience (responders) to develop particular criteria for your response sessions. You must consider where you are in the process. For example, early drafts focus more on concept, design features, and organization, while later drafts emphasize readability, usability, visual quality of the document, and final presentation. The following considerations should help you develop your own criteria as you structure your feedback sessions.

Organization/structure

How does the portfolio work as a whole? Does it have a logical, easy-to-read structure? Does it have all the elements of a book, including a table of contents, an introduction, and clearly delineated sections? How does the work flow and project a flexible structure?

Contents

Do the contents represent the range of documents in the portfolio? Is there a balance in terms of documents and sections? Are the contents presented in a logical order? How do the contents connect to your professional goals as a generalist or specialist?

Range and variety of skills represented

What skills are represented in the portfolio? Is a wide variety of skills demonstrated? Is it easy to find the skills through clear labels and context statements?

Consistency/theme

How is the theme or metaphor used throughout the portfolio? Does the portfolio have a consistent design? What similarities between the paper and electronic portfolios establish consistency between them?

Document design

How effective are the documents in terms of design, layout, and visual appeal? Are the documents arranged on the page or electronically to be both attractive and accessible?

Visual/graphic elements

How do the visual elements tie the portfolio together as a whole? Are they clear and sharp in quality? Does the portfolio reflect effective document design, image quality, captions, and an overall attractive visual impression?

Technological elements (primarily for the electronic portfolio)

How does the technology contribute to or weaken the electronic portfolio? How has the designer considered elements such as navigation and technological access?

Audience

How will others interact with the portfolios? How do the portfolios reflect industry expectations and standards?

Editing

Although this should not be the main emphasis of feedback, you might proofread the whole portfolio—including individual documents—and comment on areas of grammar and correctness.

CONDUCTING A WORKSHOP SESSION

If you are in a classroom setting, you might participate in a workshop session to get feedback on your portfolios. As we said earlier, if you are not in an academic setting, you will have to hand-pick and organize feedback individually. Whether you are in a classroom setting or on your own, you must establish criteria and guidelines for response. Practitioners use workshops in many writing and communications classes. As suggested in the previous sections, you should get feedback during all phases of this project. This section focuses on how to set up and participate in a workshop session on mature drafts of your portfolios for final revision.

We find workshops beneficial because they support a collaborative environment and encourage social learning. They are particularly useful when participants are all working toward a similar goal and have common knowledge of industry expectations and technology standards. You might even invite professionals to participate in addition to peer group respondents. Consider starting with a mock session in which you respond to an online portfolio. This will allow you to practice and analyze your responses to others in terms of both focus and mode. Once you have completed the mock session, reflect back on its purposes and results. The workshop should help its members move toward substantive revision. It is not necessarily the goal of the workshop to arrive at a solid consensus. Instead, think about collecting multiple perspectives and looking for patterns and issues to consider as you revise.

Begin by establishing criteria for the session (use Exercises 7.2 and 7.3 as a starting point). This might require some outside research to determine community expectations and effective models. It is important that as you develop criteria, the group talks about this as a way of structuring guidelines. In the assignments at the end of this chapter, you will find a sample of a response grid sheet and guidelines for a portfolio workshop session. Use these sheets or design ones that fit your particular community and industry expectations.

EXERCISE 7.2 DEVELOPING PORTFOLIO CRITERIA

For this exercise, return to the online portfolio Web sites you analyzed in Chapter 1 (if you did not complete this analysis, you can search new sites at this time). Use these sites to create a list of criteria that you feel are important and productive, focusing on both strengths and weaknesses. Generate a list of issues that you consider vital and try to define each one through ideas and questions for a potential group of responders.

EXERCISE 7.3 DEVELOPING GROUP PORTFOLIO CRITERIA

Form groups of four or five and have the members share criteria lists. Look for patterns and similarities along with differences. As a whole group, work to categorize and simplify the criteria and create a communal peer response sheet that your audience can use when providing feedback on your portfolios.

During the presentation of your work, discuss your intentions and design, displaying both the paper and electronic portfolios (it is best to project your electronic portfolio on a large screen for easy viewing). Go through all the major features, including contents, theme, and design. During the presentation, responders should take notes and write specific comments using agreed-upon criteria.

When the presentation is finished, respondents should review their comments and choose several for the full group discussion. It is generally best to have a facilitator or moderator to run the session and monitor the order and time. Respondents should work to create a discussion that is issue based rather than list based. In other words, rather than considering several issues at once, respondents should bring up a particular issue for discussion by the whole group. For example, a respondent might turn the group's attention to a part of the document—such as the introduction—or to an element such as theme, design, or color. As stated before, it is more important to encourage discussion and multiple perspectives than to reach a consensus.

As people respond to your portfolios, you might find it a bit uncomfortable to listen and not defend your choices. Try to resist an immediate reaction to their comments. You will have time later to process them and make your own choices as you revise. Hopefully you are participating in a community that has established trust and authority along with carefully designed guidelines. Although it might be difficult to hear others talk about your work, consider Trina's perspective as you participate in your workshop session:

> I tend to take criticism personally. I know it is wrong, and anyone who criticizes my writing is not really going after me personally. I must say, as a working writer, I am learning to handle the criticism as just a matter of difference.

She understands that criticism is "just a matter of difference" and that you can grow by considering multiple perspectives of your work. Try to deemphasize the feeling that the comments are personally motivated and instead view them as helpful suggestions designed to produce stronger work.

PROCESSING AND IMPLEMENTING FEEDBACK

There is no one way to interpret your documents; this is why you expose them to multiple readers with different interpretive lenses and experiences. The workshop session should not be perceived as overtly critical. In fact, you might see it as Norma does, saying, "After sharing my ideas with the class, they gave me the confidence to believe that my ideas were good." You also might find that you have serious disagreement on a particular item or element. Ultimately, it is not your job to accept all of the advice gathered in your workshop sessions. You are not trying to please everyone and should not feel manipulated by these comments. Upon revision, you still maintain control and ownership of your final work. However, the workshop session will help you to consider possibilities for revision. For example, here are several of the revision suggestions brought to Nanette's attention during a feedback session with her mentor:

* The theme elements and color scheme are strong and tell a great story about Nanette's own academic journey. I would only suggest that the "communication specialist" page bring the theme full circle and the portfolio to a nice close with elements of "the journey," "paths taken," "combining skills," etc.
* Remove the "class project" language in the descriptions to keep a more skilled, professional tone.
* The amount and variety of the content in the hard-copy portfolio felt like the perfect amount as I flipped through.
* Strengthen navigation in the online portfolio.

Nanette must process and decide which of these suggestions (along with others from her classmates) she will implement. After the group workshops of your portfolios, you will also be faced with the task of revising. You should have collected many different perspectives and issues to consider as you revise. It is your job as a writer to consider the suggestions and make choices as you rewrite and revisit design issues. In your revision, consider both global and local changes in relation to your ideas and structure. Refer to the peer response criteria you developed to determine directions for your revision.

The best way to start is to categorize all of the comments you received. This is often difficult. We find it helpful to transfer the comments from multiple forms or sheets to a single location. You might also use the response criteria to create subcategories. As you review the comments, there will be some that you immediately discard for some reason. Consider time, technical knowledge, and design abilities when deciding which revisions make sense. Some respondents might suggest ideas that sound great but are unrealistic because they require skills or materials that you might not have, such as flash animation or original illustrations. Other suggestions might be interesting but could take you away from your own purposes and intentions.

Once you have discarded suggestions that you don't want to use, you can focus on those that are left. Pay particular attention to ideas that are mentioned several times. Even though you might find disparity in the nature of the comments, when you have multiple responses to an issue, they should motivate you to at least consider the suggestions. Focus on areas of strength as well as weakness; you can learn a lot from what people admire and emphasize or duplicate that feature throughout the portfolio. You might prioritize the comments according to their importance and practicality. Create one list that includes your top priorities and another that you might address if you have time.

Once you have decided on the issues to focus on in your revision, go back to your portfolios and start to make changes. Of course, if you can get additional feedback on successive drafts for quality control, this is extremely helpful. The more feedback you get as you revise, the better.

EDITING

After completing your final revisions, it is essential that you carefully proofread the portfolios to eliminate all superficial errors. There is nothing as discouraging or embarrassing as a portfolio that does not reflect the careful eye of a trained communicator. Although we all make mistakes, it is not uncommon for an employer to immediately cast aside a portfolio that contains such oversights. Carefully **edit** your portfolios to eliminate errors in typing, spelling, documentation, and grammar. It is time to shape and refine your text for professional presentation.

Carefully reread your whole portfolio and individual documents. You might try reading them aloud or looking at them in a different order or format. However, as we noted earlier, it is often difficult to edit your own work. Over time, you have become so close to it that you might overlook obvious mistakes. At this point, it is best to seek the help of an objective, trained reader, such as a colleague or classmate, to proofread and edit your portfolio. Make sure that you choose someone who is familiar with the grammatical and mechanical conventions of your field. Ask this person to also double check your citations and instructions.

SUMMARY

Your portfolios are constructed for particular audiences and purposes. Chapter 7 helps you to connect with your audience as they review your portfolios, describing guidelines and processes for eliciting feedback throughout the portfolio-building process. The tips provided in this chapter should help you to draft your portfolios and create response criteria, as well as conduct a response workshop session and implement feedback. Brian D. sums it up nicely with his closing comment: "Probably the most important idea I learned in this class is that you should always talk a design through with at least one other person."

ASSIGNMENTS

Assignment 1: *Conducting a Peer Response Workshop*

The following guidelines and forms summarize the chapter discussion to help you set up a peer response workshop for your portfolios.

PORTFOLIO WORKSHOP GUIDELINES

The goal of these workshops is to provide productive suggestions for revision. It is important for the participants to show mutual respect and a genuine desire to help one another. Harsh or judgmental comments are unacceptable. Although we are looking for solid, constructive feedback, we hope that you will consider the tone, purpose, and audience in providing comments (both written and verbal).

- **As a group,** you will need to choose the order of the presentations, as well as a moderator for each workshop session (every person should get a chance to moderate a session).
- **As the moderator or facilitator,** it is your job to promote the discussion and to be a particularly close reader of the portfolios. It is also your responsibility to keep time. If the discussion starts to lag, bring up new points for discussion and help the group summarize the suggestions at the end of the session. You should also take notes during the session.
- **As a responder,** it is your job to come to the session ready to focus and give thoughtful written and verbal feedback to your classmates. You should also pay attention to the issues discussed, as they most likely represent universal issues that you might reflect upon as you revise your own work.
- **As the author/designer,** it is your job to listen to the group's comments without reacting. You don't necessarily have to take all the advice, but the discussion and written comments should help you reconsider your ideas and choices and the ways they are interpreted by your audience. You should be prepared with questions that will enable you to get the most out of your response session. Remember, this is a great opportunity for you to get ideas for the revision of your portfolio. Since there will not be time to consider all the verbal feedback suggestions, you will also have the Portfolio Feedback sheet to consider in making revisions.

Workshop Structure

The workshop session should run as follows (30 minutes):

15 Minutes: Visual Presentation of Portfolios

- **Presenter:** In a short presentation, discuss and display your portfolios, including samples of introductory sections, individual selections, and overall layout and design. The electronic version should be linked and active.
- **Responders:** Take notes and record impressions during the portfolio presentation (Portfolio Feedback sheet). Take a couple of minutes after the presentation to transform notes into concrete revision suggestions.

15 Minutes: Revision Feedback Discussion

- Start the discussion by talking about the strengths of the work.
- Then discuss revision suggestions—areas to develop further and other issues.
- The author asks final questions for clarification.
- Submit all written feedback to the author (Portfolio Feedback sheet).

Note: All discussions should focus on issues that are considered by the whole group. Talk through the responses and generate multiple perspectives.

Assignment 2: *Workshop Response Grid*

The following grid is a model of a form that you might use for the workshop response. Modify or use it to respond to your classmates' work during workshop sessions. Return copies to the author/designer upon completion of the workshop so that they can consider the responses and implement revision suggestions.

Name of Presenter _____

Name of Responder _____

Portfolio Feedback: Your feedback will help your classmates revise their portfolios. Please circle the number that best represents your initial impression and follow up with written notes to justify your rating.

Scale: 1 = Excellent   2 = Good   3 = Average   4 = Weak   5 = Poor

<u>Paper Portfolio</u>

Organization/Structure   1 2 3 4 5

Comments:

Content   1 2 3 4 5

Comments:

Range and Variety of Skills Represented   1 2 3 4 5

Comments:

Consistency/Theme   1 2 3 4 5

Comments:

Visual/Graphic Elements   1 2 3 4 5

Comments:

Overall Impression   1 2 3 4 5

Comments:

<u>Electronic Portfolio</u>

Home Page   1 2 3 4 5

Comments:

Visual/Graphic Elements   1 2 3 4 5

Comments:

Navigational Structure   1 2 3 4 5

Comments:

Coordination/Consistency with Paper Portfolio   1 2 3 4 5

Comments:

Overall Impression   1 2 3 4 5

Comments:

Suggestions for Revision

At the end of the presentation, take a couple of minutes to provide concrete suggestions for revision. Use the ratings and comment notes to suggest points for revision. Choose two or three issues (grounded by specific text examples) to discuss with the group.

REFERENCES

Barnum, Carol. *Usability Testing and Research*. New York: Longman, 2002.

Straub, Richard and Ronald F. Lunsford. *Twelve Readers Reading: Responding to College Student Writing*. Creskill, NJ: Hampton Press, 1995.

8 Portfolios and the Job Search— Getting Prepared

I want to market myself as a versatile technical and professional communicator. I want people to see that I can adapt to different situations. Norma

INTRODUCTION

Chapters 1 through 7 have guided you through the portfolio-building process, providing specific guidelines on how to select, design, create, and revise the documents to be included in your portfolios. Now you are ready to plan your strategy for using the portfolios in your job search. Before taking the portfolios to a job interview, you will need to research the type of job you want and create or update your employment documents. Chapter 8 presents guidelines on how to do the following:

* Developing a marketing plan for your job search
* Preparing or updating your employment documents

Chapter 9 continues the discussion by focusing on the job interview, including how to use your portfolio during the interview.

DEVELOPING A MARKETING PLAN

In today's job market, the name of the game is change. A typical worker today can expect to have as many as 12–15 different employers during his or her work life, working just long enough to complete a project or two (Bolles 1999, 141). Conlin uses the term "free agents" to describe this new workforce and estimates that roughly 41 percent of the U.S workforce will be free agents by 2010 (Conlin, 2000, 169–170). No matter what your major, you can expect multiple employers and jobs during your professional career (Lannon 2003, 434). This is particularly true for technical and professional communicators.

To prepare for the changes you will encounter in the workplace, it is important to develop a marketing plan for your job search. If you are about to begin your career as a professional communicator, your marketing plan may be based on the following activities:

* Identifying your strengths and weaknesses
* Interviewing a professional communicator
* Finding a mentor
* Identifying your market
* Researching and collecting information on companies that employ technical and/or professional communicators
* Learning the language of potential employers
* Finding information on companies in your field
* Conducting an Internet job search
* Creating a network of contacts

Identifying your strengths and weaknesses

You probably did a self-assessment when you started developing the portfolio plan. However, you must now consider your strengths and weaknesses from a slightly different perspective, relating them to the job market you are about to enter. A well-known syndicated cartoonist offered the following advice to a graduating college class: "Identify what you want to do in life and then find someone to pay you to do it." That advice is still valuable today, and a self-assessment will help you identify what career path in technical and professional communication you may want to follow. Two of the more popular books that can help you identify strengths and weaknesses are Richard Bolles' *What Color Is Your Parachute?*, now in its 33rd edition, and Janet Van Wicklen's *The Tech Writer's Survival Guide*. There are many Web sites that offer personality profile tests, and several are discussed in Chapter 2. Some of these tests are free; other sites offer a free introductory version but charge a fee if you want to take the complete test. You can find these sites by doing a keyword search using your favorite search engine. Another very effective way to identify your strengths and weaknesses is to do the activities in Exercise 8.1.

EXERCISE 8.1 FOUR WAYS TO UNCOVER YOUR ABILITIES

1. List five of your achievements in order of importance.
2. Identify the skills that contributed to each achievement.
3. List what you liked best in work you've done before.
4. List examples of your soft skills, that is, skills that can't be quantified or easily measured, such as interpersonal skills, ability to work on project teams, and other people skills that employers mention in their job ads.

After completing your self-assessment, you may want to return to the professional identity work done in Chapter 2 with a new eye toward revising it for your job search. In her revised personality profile, Miranda, a senior graduating with a degree in technical and professional communication, matches her willingness to take calculated risks and a desire to share what she's learned with her decision to pursue freelance work. Miranda makes the following observations regarding this decision:

> It hits me as I read about how freelance work is actually more secure and how freelance writers can market themselves while supplementing their income. I think to myself—I can do this! And I think I can do this because I'm in control. I don't mind risk as long as I can manage it, understand it. I think this confidence that I can understand and manage risk comes from my love of planning. Although I was stressed all semester long over my portfolio, I felt okay with the progress because everything was planned. I knew where I was going and how I was getting there.

Michel, on the other hand, describes her dream job as that of a writer/editor for a large government agency or pharmaceutical company. Her decision is based on a different set of personal preferences:

> The best job I could imagine would be to work for the CDC or certain types of pharmaceutical companies as a writer/editor. I think I'd prefer writing or editing disease/drug information that goes out to the public. I tend to have a reader-friendly style, which is important for transmitting complicated medical information to the public. As long as I make enough money to live on, salary isn't a big issue. With the government, I'd have great benefits, job security, and stability.

In his personality profile, Brian W., a senior graduating with a degree in technical and professional communication, links his love of graphics and technology to a career in professional communication that stresses video production and multimedia:

I see myself fitting into the technical communication profession through the integration of technology and art. This could be through video production or professional presentations. I don't seem to enjoy writing that much. I could write technology reviews, and I would probably enjoy that, but I don't foresee myself writing manuals and being happy about it. I have the skills and the interests to go on and either start my own multimedia business, contract myself out to other companies, or begin working full time in a company doing creative multimedia work. It [my career choice] will give me control over what kind of work I do, and if I don't deem it to be enjoyable, I can move on to other work that is.

Interviewing a professional communicator

Particularly if you have little or no work experience in technical or professional communication, you should interview someone in that profession. Interviewing a professional communicator will help you learn the realities of the workplace. The person you interview can provide tips on how to conduct a successful job search and how to match your strengths with a specific job description, among other things. This person might even agree to become a mentor, advising you as you create and later revise your resume and portfolio.

Finding a mentor

If you are putting together a professional portfolio, your instructor may assign you a mentor. This person can be a wonderful resource to call on when you plan, select, and organize the artifacts for your portfolio. If you are an entry-level technical writer, you may want to ask a senior writer to help you select your best work for the portfolio. When Nanette was selecting portfolio projects for her senior capstone course in her major, she met with her mentor, who suggested that she add more "spark" to her portfolio. Nanette showed him some street sign images that she was thinking about using as graphic symbols of the different career paths open to her as a professional communicator. He suggested that she use these signs as a theme for her portfolio. Brian W. also received helpful tips from his mentor, who suggested that he add more multimedia projects to his portfolio to emphasize his interest in finding a multimedia position. Another effective way to find a mentor is to check the Web sites of local chapters of professional communication organizations such as the Society for Technical Communication that often have mentorship programs. The career center at your school may also offer resources for seniors.

Mentors can also help you learn more about the job market and help you find a position that matches your specific interests. For example, Norma set up an interview with her mentor, a knowledge-base content developer, a relatively new area of technical communication that focuses on helping companies maximize their intellectual capital. From the interview, Norma picked up several tips on how to conduct a successful job search, including the following:

* Identify the work environment where you would be most productive (casual or formal, large or small company).
* Build a network of friends, relatives, and working professionals who can provide job leads.
* Be willing to relocate to places where the opportunities exist.
* Keep an accurate logging and tracking system on the companies you have researched.

Trina, who was interested in finding out what writers do in information technology departments, interviewed Mary, the sole technical writer in such a department. Mary provided a list of her job responsibilities, which included the following:

* Review outstanding projects and ad hoc requests for application documentation and help files.
* Periodically review training materials created for applications.
* Create user documentation from business analysis documents and screenshots to develop a Web site help system.
* Train new users.

Trina also learned that Mary's work environment was a small office cubicle where Mary worked alone on her projects.

Sarah, who is interested in a public relations and marketing career, interviewed Susan, the director of marketing at a small software company. Susan commented on how "marketing" can sometimes be a generic term for many different functions such as corporate communications, product marketing, and brand marketing. Sarah learned that brand marketing can include public relations (press releases, media relations, and speaking events), trade shows (demonstrations, signage, session presentations, and contests), corporate and product marketing documents (brochures, CDs, flash presentations, and audiovisual storyboarding and scripting), and Web site development and maintenance. Sarah also gained some valuable tips on how to market herself:

- Join the American Marketing Association.
- Network through local professional writing associations, marketing associations, alumni, and professors.
- Use corporate Web sites to research local companies that have visible marketing and public relations functions.
- Do contract work through an agency (a good way to transform a temporary or part-time position into a full-time position).

Identifying your market

Clearly, there are many methods that you can use to find a job. Table 8.1 summarizes 12 of the most common ones. As the table shows, almost half of the search methods involve some form of networking. Creating a network of potential job leads may be the most valuable strategy you develop for your job search. Networking is discussed in more detail later in this chapter.

TABLE 8.1
Most Commonly used Job Search Methods

Percent of Total Job Seekers Using the Method	Method	Effectiveness Rate*
66.0	Applied directly to employer	47.7%
50.8	Asked friends about jobs where they work	22.1%
41.8	Asked friends about jobs elsewhere	11.9%
28.4	Asked relatives about jobs where they work	19.3%
27.3	Asked relatives about jobs elsewhere	7.4%
45.9	Answered local newspaper ads	23.9%
21.0	Private employment agency	24.2%
12.5	School placement office	21.4%
15.3	Civil Service test	12.5%
10.4	Asked teacher or professor	12.1%
1.6	Placed ad in local newspaper	12.9%
6.0	Union hiring hall	22.2%

*A percentage obtained by dividing the number of job seekers who actually found work using the method by the total number of job seekers who tried to use that method, whether successfully or not.

Source: U.S. Department of Labor, Employment and Training Administration (1996), 8.

Researching and collecting information on companies that employ technical and/or professional communicators

Whether you are a student or a recent graduate, you may want to begin researching and collecting company information by searching the job databases at your university's career resources center. However, this search for technical writing and professional communication positions may turn up very few job leads. Don't despair. The job titles "technical writer" and "professional communicator" are very broad, encompassing several more specific job titles and descriptions such as those mentioned in Chapter 1. In "Online Job Searching: Clicking Your Way to Employment" (2003), Janel Bloch suggests that you search for job descriptions under a variety of terms, including the following:

* Audience
* Graphic design
* Information design
* Instructional design
* Medical communication, medical writer
* Technical communication
* Technical editor, technical editing
* Technical writer, technical writing
* Usability
* Web design

Source: Written by Janel Bloch and used with permission from INTERCOM, the Magazine of the Society for Technical Communication, Arlington, VA, U.S.A

You may also want to consider using these additional terms in your search:

* Communications specialist
* Corporate communications specialist
* Editor
* Journalist
* Knowledge manager
* Marketing publication specialist
* Multimedia specialist
* Public relations
* Translator
* User documentation specialist
* Video production
* Web usability specialist

Another potentially effective search technique, Bloch notes, is to combine job descriptions with specific software tools such as PageMaker or FrameMaker (Bloch 2003, 14). Even though technical and professional communicators refer to themselves by a variety of job titles, no matter what the title, the job outlook for technical communicators is very bright. The *U.S. Bureau of Labor Statistics Occupational Outlook Handbook* (2004–05, 1, 6) predicts a 10 to 20 percent growth rate up through 2012.

Learning the language of potential employers

You should become very familiar with the language that potential employers use in describing the positions they offer for professional communicators. Do this by beginning your job search early and by reviewing as many job advertisements on the Internet and in print as you can. The following description of an entry-level professional communication position with an instructional design company uses the terms "sales training," "customer service," and "multimedia." If you decided to apply for this position, you would probably want to use some of these terms in your letter of application and perhaps even on your resume. This is particularly true today, since so many resumes are optically scanned for terms that match the terms the employer has used in the job description. Tips on how to write a scannable resume are presented later in this chapter.

SAMPLE JOB DESCRIPTION FOR AN INSTRUCTIONAL DESIGNER POSITION

Our projects cover a wide range of content areas including sales training, software training, product knowledge, customer service, operations, logistics, performance appraisal, process and procedures, and others. We develop our training packages in print, Web-based, multimedia, video, and audio, as well as converting materials for Web distribution. We also design and produce job aids, including quick reference guides, and online help for end users.

EXERCISE 8.2 SELECTING TERMS TO USE IN A SCANNABLE RESUME

Examine the following job description and write down five terms that you would consider including in a scannable version of your resume. Be prepared to justify your choices.

Company:	Life Blood Community Center
Job Title:	Technical Writer, Full Time
Job Description:	This person is responsible for writing and editing standard operating procedures, user manuals, HTML documents, and documents for internal use, customer/donor reference, or publication.
Education:	Bachelor's degree in technical communication, English, or a related field
Experience:	0–3 years of technical writing and editing experiences
Requirements:	Excellent oral and written communication skills, strong understanding of the English language, basic computer knowledge including Word, ability to understand and follow instructions and procedures, ability to prepare and print routine correspondence and other basic written material, ability to file and maintain records; basic HTML skills desired, but not required.

Finding information on companies in your field

Time after time, employers note how impressed they are with job candidates who have researched their companies extensively and can talk in detail about the company's products and services at the job interview. One hiring manager at a software company observes that a job candidate should have two to three specific questions about his company's products and services to ask at an interview. Surveys conducted by the National Association of Colleges and Employers indicate that job candidates who have done their homework and who can discuss how their experiences and qualifications match the company's needs are typically the ones who get hired (Crowther 2000, 33). Because many professional communicators are hired by relatively small firms, make sure that you research small companies as well as large ones.

Table 8.2 identifies some key sources that you may want to consult, along with questions that the sources may help you answer.

You may also want to use your favorite search engine to locate specific company Web sites. These sites will often provide company information on products, services, employment opportunities, company history, and research and development activities. You can also contact your local chamber of commerce and your better business bureau, and use CD-ROMs available in your school's library.

TABLE 8.2
General Job Search Sources

Source	Relevant Question
Annual report (often available at company Web sites, your career center, or in your library)	What are the company's products and services? Are stockholders making a profit?
Web sites or media kits (available from company's public relations office)	What can you learn about the firm's corporate culture?
Personnel manuals and other policy guidelines	How committed is the firm to training? What are the benefits and retirement programs? What are its customary career paths?
Graduates of your college or university now working with this firm	What sort of reputation does your school have among decision-makers at the company?
Business sections of newspapers and magazines	What kind of news gets generated about the company?
Professional organizations or associations	Is the company active within its profession?
Stock reports	Is the firm making money? How has the company performed during the last 5 years?
Accrediting agencies or organizations	How has the company fared during peer evaluations?
Former employees	Why have people left the company?
Current employees	What do employees like or dislike about the company? Why do they stay?

Source: Adapted from Pfeiffer, William Sanborn, *Technical Writing: A Practical Approach,* 5th Edition, © 2003. Reprinted by permission of Pearson Education, Inc., Upper Saddle River, NJ.

EXERCISE 8.3 ANALYZING COMPANY WEB SITES

Find and then analyze three company Web sites for information on the following:

- Products and/or services
- Employment opportunities
- Recent research and development activities

Write your analysis (approximately one paragraph per company).

Conducting an Internet job search

For technical and professional communicators looking for work, the Internet is a very valuable resource. You may recall from Chapter 1 that Brian D., a recent graduate of a professional communication program, received an interview and a job offer based exclusively on his Internet resume and portfolio. Remember, however, that an Internet search should be only one of your job-search strategies, and you need to be an educated user if you decide to post your resume to any of the job-search sites providing that service. Most popular job-search sites provide different types of privacy protection for users. However, as Janel Bloch notes in "Online Job Searching," none of these safeguards are foolproof (Bloch 2003, 12).

The most useful Internet job-search sites will be one of these types:

* General search sites
* Job search sites related to technical and professional communication
* Professional association sites

Most general search sites have extensive databases of job listings for many different types of professional job seekers. The more popular sites provide an "Advice and Resources" link to many helpful job-related articles (Bloch 2003, 13). Other general search sites provide links to articles on self-assessment, resumes, and interviewing tips.

Table 8.3 provides a brief summary of some of the more popular general job search sites that you may want to use to begin your Internet search. Appendix A lists the URLs for these sites.

Table 8.4 lists some of the more popular job-search sites for technical and professional communication. The URLs for these sites are listed in Appendix A.

TABLE 8.3
Sample Career Resource Websites

Web Site	Description
America's Job Bank	Provides a computerized network that links 1,800 state employment services offices. This site has been selected by several Internet rating services as one of the top Internet sites.
Avue Central	A free federal employment service that lists over 35,000 government jobs.
Careerbuilder.com	Claims to be the nation's largest employment network. Includes a host of job-related links. Also includes personal job searches and job tips.
Careers at Yahoo!	Allows you to browse jobs by subject, post your resume, and find a list of the most popular jobs. Also provides a career toolkit with resume tools, job alerts, and a salary wizard.
FlipDog.com	Claims to deliver the largest number of jobs on the Internet, including those of small companies and public and private companies. Services are free to the job seeker. Provides services for posting resumes and applying for jobs.
Hotjobs.com	Similar to Careerbuilder.com. Has a large database of jobs to search from, a very detailed search engine, and a career tools section where you can find articles and information about your industry.
Job Bank USA	Features U.S. and international job listings. A special feature lets users search any of the 12 largest Web job sites
Monster.com	Includes a job search feature, a resume-posting service, and a list of featured employers. Has a career center providing career advice and community discussions.
Monster TRAK	Claims to be the number one Web site for alumni and students looking for full- or part-time employment. Claims to be used by more than 600,000 employers. Provides extensive job search features with links to college and university career services. Formerly known as Job TRAK.
The Careers Organization	Provides information on jobs, employers, education and learning, and career services.

TABLE 8.4
Sample Technical and Professional Communication Job Search Sites

Web Site	Description
Computerjobs.com	Has an entire section devoted to technical writing. You can search for jobs based on your criteria, such as job title.
ELance	Designed to help freelancers find projects that match their skills. Offers opportunities to search for and bid on projects.
Newbie Tech Writer	Provides links to a variety of job-search Web sites for new technical writers.
Technicalwriterjobs.com	Divided into several areas including "Find Technical Writer Jobs," employers, recruitment firms, and advertising agencies.
TechWriting at About.com	Provides information about freelance writing, job-search information, contracts, grant writing, journalism, style guides, and professional groups.
Techwritingjobs.com	Set up specifically for technical writers. Allows you to search for jobs and offers a bulletin board of job fairs.
The Write Jobs	Posts job opportunities for writers. Positions include journalism, editing, writing (freelance and staff), and technical writing.
Webdeveloper.com	Provides a variety of resources relating to Web site development. Includes a technical job center service.
Writerfind.com	Describes itself as a marketplace for professional writers and employers who are looking for freelance and telecommuting jobs or posting such jobs. Job postings are international and include technical writers, instructional writers, and scriptwriters.

Table 8.5 provides a short description of some of the major professional associations for writers and communication specialists. Again, the URLs are listed in Appendix A.

Remember that these lists of sites comprise only the tip of the iceberg.

Creating a network of contacts

The first networking tip that you should put into practice is to tell everyone you know that you are looking for a job. The knowledge-base content developer mentioned earlier in this chapter notes that almost every job she's ever found involved networking. For most technical and professional communicators, networking is a survival skill because most of these positions are not found through traditional sources such as your university's career center. Having your own Web site with a link to employment documents is one effective way to network. In short, as Van Wicklen notes in *The Tech Writer's Survival Guide*, networking involves the practice of establishing friendly contacts with other professionals to share work-related information (Van Wicklen 2001, 49)

How to network. Remember that when you are networking, you are marketing yourself. The networking tips in Figure 8.1 should help you do so effectively. In addition to contacting the professional organizations listed in Table 8.5, you may want to extend your networking by finding out more about the following professional communication organizations, whose URLs appear in Appendix A:

- International Public Relations Association
- Society of Environmental Journalists
- Society of Publication Designers

TABLE 8.5
Selected Professional Associations

Professional Association	Description
American Marketing Association	A major association of marketing professionals with a membership of 38,000. Includes a career center where you can post resumes, search for jobs, and find salary information.
American Medical Writers Association	The major professional association for medical writers. Its mission is to promote excellence in biomedical communication. Provides job market information including a job-search feature.
American Society for Training and Development	The major professional association for trainers. Offers a career center with job listings and a resume bank, as well as a comprehensive training literature database.
American Society of Journalists and Authors	Provides professional freelance writers with a variety of services including rates, contracts, and conferences. Includes a writer referral service that provides access to jobs and project leads.
Association of Health Care Journalists	A nonprofit organization advancing the public's understanding of health care issues. Promotes highest standards in reporting, writing about, and editing health care journalism. Includes job postings.
Graphic Artists Guild	Provides a jobline news service listing freelance and staff employment opportunities in the graphic arts industry.
International Association of Business Communicators	Includes a job center for both job seekers and employers where jobs are listed and resumes can be posted and viewed.
International Webmasters Association	Provides professional advancement opportunities for individuals pursuing a Web career. Offers a variety of services including a jobs and resume resource for Web and information technology industries.
National Association of Science Writers	The major professional organization for science writers that provides a link to job services for employers and members. Also provides a freelance site.
Society for Technical Communication	The major professional organization for technical writers, technical editors, information architects, usability specialists, and other technical communication professionals worldwide. Provides a variety of links including links to local chapters.
The Usability Professionals' Association (UPA)	Supports professionals who advance the development of usable products and who act as advocates for users. Provides links to a consultants' directory and job bank.

You should also develop a tracking system to organize your networking and job-search activities. You can do this by creating an electronic spreadsheet with detailed information on who you contacted, how, and when. In addition, summarize what was said and what action was taken, if any. Finally, note what the next step will be and who will take it. You may want to print out a copy of your recent job-search activities so that you can check it often to avoid missing deadlines.

---

1. Develop a list of contacts including business associates, classmates, teachers, and friends. Give each of them your business card with the URL of your Web portfolio.

2. Target companies that you would like to work for well in advance of your graduation and before you need another job.

3. Learn about your target companies' products/services and the types of jobs they have in professional communication. Try to find out about their hiring procedure.

4. Prepare a call list to work from daily and force yourself to make a certain number of calls per day. Say that you are calling for advice on how to enter the field of professional communication.

5. Try to get over the feeling that you are bothering others. Instead, think of your questions as prompting them to talk about a mutual interest: professional communication.

6. Subscribe to related industry trade journals and newsletters.

7. Subscribe to list serves and e-mail networking services that pertain to professional communication.

8. Take seminars and courses.

9. Keep your networking conversation brief, focused, and courteous.

10. Become active in your professional organizations (see Table 8.5), including any special interest groups (SIGs) or local chapters in your city or town.

11. Use any mentoring services provided by your local professional communication organization. Often you will find a link on that organization's Web site.

12. Stay in the loop. Be persistent but tactful. Call back in 5 or 6 weeks if you don't hear from someone.

13. Remember to thank those who helped you.

FIGURE 8.1
Thirteen Key Networking Tips

PREPARING OR UPDATING YOUR EMPLOYMENT DOCUMENTS

At a minimum, your employment documents should include the following:

* Current resume
* Letter of application/cover letter
* References
* Portfolio

Preparing a current resume

Whether you are preparing for your first job as a professional communicator or are already employed in this position, you will need to update your resume to include your current skills and abilities. The purpose of your resume and cover letter is to get you interviews. You can do this by organizing the information on your resume so that you grab the attention of a potential employer in 20 to 30 seconds. Within this time period, most hiring managers will have made a decision about pursuing your application.

Features of the standard resume

The vast majority of resumes will have the following features:

* Contact information
* Career or professional objective

* Education summary
* Employment summary
* Special skills, awards, certifications, professional affiliations, languages
* References and portfolio

If you are applying for your first job as a professional communicator, you should probably emphasize your education (including projects you created). As you acquire employment experience in professional communication, your resume will emphasize your workplace experience.

Contact information.   Provide your full name, mailing address with zip code, e-mail address, and phone number(s). If you have voice mail, make sure that your message sounds professional. If your e-mail address is comical or cute, change it or open a new e-mail account for your job search. If you include both a school address and a home address on your resume, make it clear which one is your primary address.

Career or professional objective.   For many job seekers, the career objective or professional objective is the most difficult part of the resume to write. It should be specific to add focus to the rest of the resume but not so specific that it might eliminate you from positions for which you are qualified. Try to avoid vague statements such as "an entry-level professional writing position where I can apply my education and experience." Instead, focus on a set of specific skills that you have already identified in your work in Chapter 2 on establishing a professional identity. You may also want to include both immediate and long-term objectives. Figure 8.2 provides two examples of effective career objectives.

EXERCISE 8.4 WRITING CAREER OBJECTIVES
Write two types of career objectives, one a skills objective and the other a short-term and a long-term objective.

Education summary.   As noted earlier, a good practice is to place your education summary before your employment summary if your education has provided most of your marketable skills, projects, and software knowledge. Use a reverse chronological order, listing your most recent or current school first and then working backward, listing the other schools where you have received degrees (see Figure 8.4). Include the exact name of your major and of any minors or certificates that you may have completed or will complete before you graduate. List relevant courses, projects, and software tools that clearly match the job you seek. Finally, you might include your grade point average if it distinguishes you from other students in your major. Your adviser or a teacher in your major subject area will tell you if your grade point average places you in the top 25 percent of your class. If you paid for part or all of your education by working full-time or part-time, it never hurts to note the percentage of your tuition earned through working.

Example of a skills objective

Multimedia specialist for a company that develops Web sites and multimedia training

Example of an immediate and a long-term objective

Immediate objective: Entry-level technical writer with a software firm

Long-term objective: Progression to project manager with responsibilities for scheduling document development and for hiring new writers

---

FIGURE 8.2
Sample Career Objectives

Employment summary. If your employment record is a bigger selling point than your education, present it first. Begin with your most recent position and work backward. Include dates of employment and names of employers, noting whether the position was full-time or part-time. If you worked part-time, include the number of hours per week. Give your job title with a short description of your responsibilities. If you were promoted, note it. If your employment history is largely unrelated to professional communication, focus more on soft skills (organizational, interpersonal, leadership) that you have honed—for example, as an assistant manager at a fast-food restaurant. In that case, you may want to arrange your resume using the combination format (see Figure 8.6). If you have military experience, note it here. If you are willing to relocate, state that too.

Special skills, awards, certifications, professional affiliations, languages. List any special skills that pertain to the type of position you are seeking. In particular, note relevant software skills—especially if you have used that software to create a piece for your portfolio. If you belong to a professional organization, note it here along with the membership dates. Mention your membership in campus organizations, civic groups, or volunteer organizations, noting any leadership positions you may have held. If you know a foreign language, make sure you mention it. In addition, list any certifications that you have, particularly if they indicate marketable skills. Finally, include any awards that you have received, such as being selected for "Who's Who Among Students."

References and portfolio. There are two methods for including references on a resume. The first, and more popular, approach is to say "References are available upon request." If you choose this method, make sure that your reference letters are available when an employer requests them. The second method is to list three or four individuals who have agreed beforehand to write strong letters of recommendation. Make sure your references are aware that they are listed on your resume and that they might be contacted when you begin your job search. Avoid asking people who don't know you well or who might be less than impressed with your work. Letters from these individuals are likely to be too general to be persuasive or might even damn you with faint praise. Select individuals who know you well so that they can discuss your character traits and job-related skills in detail.

You may want to give your references an updated resume beforehand so that they can refer to the same skill sets and work experience that a potential employer will be reviewing. Detailed letters of recommendation are time-consuming to write, so give your references enough lead time to prepare them. Finally, make sure that your resume indicates that you have a portfolio. If your portfolio is Web-hosted, provide the URL. If it is not Web-hosted, you can say, "Portfolio is available upon request." Make sure that several CD versions of your portfolio are ready to send should a potential employer request one.

Double check everything

Carefully proofread every writing sample you have included before the interview. Check all the links in your electronic portfolio to make sure that they work. Ask the interviewer what type of computer, operating system, and memory capacity he or she has so that your electronic portfolio will run in full version on that computer. Make sure that you have multiple resumes so that you can leave copies behind.

Tips for designing an effective resume

Figure 8.3 lists 10 tips that will help you design an effective resume.

Ordinarily, it is a good idea to have your resume available in two different styles so that you can select the one that best addresses the qualifications in the job ad. A scannable version of your resume should also be available to send electronically. The styles are as follows:

* Chronological style
* Functional style
* Combination style
* Scannable style

Chronological resume.   The chronological resume lists your employment history with the most recent job information first. This resume style has several advantages. It is the easiest to prepare, since its content is arranged by dates, companies, and titles. It emphasizes your steady employment record. Professional interviewers are very familiar with the chronological resume, and it provides a guide for discussing work experience. However, the chronological resume also has drawbacks. It reveals gaps in employment and can put an undesired emphasis on unrelated jobs, summer jobs, or part-time jobs. It can also deemphasize skill areas unless these areas are reflected in the most current job. Figure 8.4 shows the chronological resume of a hypothetical student named Frank, whose work experience is presented from latest to earliest. Frank also mentions relevant courses and includes a reference to his portfolio.

Functional resume.   The functional resume works well if you want to summarize skills and/or work experience developed over several years. Instead of presenting a list of jobs and dates, the functional resume allows you to highlight those duties and skill sets that best match the job description. For example, you may have

---

1. Limit your resume to one page unless you have extensive job-related skills or experience. Use at least a 10-point font.

2. Be positive. Focus on accomplishments and strengths.

3. Do not mention salary requirements. Save that discussion for the interview.

4. Use formatting techniques like boldface, bullet lists, parallel construction, and white space to make your resume easy to review.

5. Use phrases rather than complete sentences.

6. Rely on action verbs to describe qualifications and skills.

   Some verbs describing communication skills

authored	communicated	created	designed
edited	marketed	planned	presented
produced	published	revised	trained

   Some verbs describing management skills

administered	advised	coordinated	directed
evaluated	implemented	improved	supervised

   Some verbs describing research skills

administered	analyzed	collated	composed
coordinated	designed	diagnosed	distributed
evaluated	interviewed	investigated	organized
summarized	surveyed		

7. Be specific. Design your resume with particular job requirements in mind. Use the language of potential employers by reviewing job ads. Focus on technical and software skills.

8. Include quantifiable job contributions (e.g., "Modified our existing documentation process, which saved our department $75,000 a year").

9. Use a high-quality white or off-white bond paper to print your resume (at least 20 pound bond and 25 percent cotton content).

10. Keep in mind that your resume is a writing sample and proofread it. Have a friend proofread it and then proofread it yourself one more time.

FIGURE 8.3
Ten Resume Tips

held writing positions at three different companies during the past 3 years. Instead of focusing on the dates and job titles, you may want to provide a career summary that describes the various document design skills and software tools you learned to use on these jobs. Even if your work experience is not extensive but you do have a marketable set of skills, you may want to use the functional resume because it deemphasizes specific

---

Frank Prater

1390 Trailedge Way
Malone, NY 16789
315.723.0391 (home)
315.534.8932 (cell)
fprater@earthlink.net

Objective   Technical writing position using my writing skills, document design skills, and strong technical background

Education
2001–present   **Bachelor of Science in Technical and Professional Communication with a minor in Computer Science (expected May 2004)**
Malone University, Malone, NY
GPA 3.89 (out of 4.0)

Major courses include: Advanced Grammar and Editing, Journalism, Manuals, Proposal Writing, Public Speaking, Rhetoric, Science Writing, Technical training and coursework in document design and multimedia design

Minor courses include: C++ Programming, Computer Data Structures, Database Systems, Java Programming

1997–2001   Syracuse Institute of Technology
Studied Chemical Engineering
Coursework included: 33 semester hours in German including Advanced Stylistics, Conversation and Composition, German Business Communication, German Science and Technology, German Film and Literature

Experience
June 2002–   **Database Developer, Malone Sweeteners**
Responsibilities include:
• Design and implement a database to track customers and products
• Complete data entry of existing customers, contacts, and products
• Develop a user's manual for the database

1999–2001   **Assistant Manager, Super Video**

1997–1999   **Staff Lead, MegaPlex Cinema**
• Promoted from Floor Staff to Staff Lead in June 1998
• Developed management and leadership skills

Computer Skills   Microsoft Office Suite (Word, Excel, PowerPoint, Access), Adobe FrameMaker, Adobe PageMaker, Adobe Photoshop, HTML

Achievements   **Outstanding Technical and Professional Communication Senior, 2003**
2nd Degree Taekwondo Black Belt

Portfolio   Professional portfolio available at interview
Electronic portfolio available upon request

References   Available upon request

FIGURE 8.4
Sample Chronological Resume

job titles and dates. Figure 8.5 is an example of a functional resume. Mark's background summary highlights his training and instructional design experience so that an employer has a clear picture of his employment strengths before reading the rest of his resume.

Combination resume.   The combination resume uses characteristics of both the chronological and functional resumes, providing versatility and allowing you to match your skill sets to specific job requirements. Another advantage of this resume is that it focuses on activities and skills rather than dates, thus deemphasizing employment gaps. The combination resume does have drawbacks, however. Its format may be less familiar to recruiters, and important dates may be difficult to find. It can also be longer than the chronological resume. Figure 8.6 is Wilda's

	Mark DesRoches **5218 Meadowdale Court** **San Francisco, California 94102** **(913) 298-2107** mdesroches@yahoo.com
Objective	Position as an instructional designer developing Web-based training and online help systems for a variety of clients
Background Summary	Five years of instructional design experience including the development of Web-based training programs for Fortune 500 companies, writing training materials for instructors and participants, and contributing to the development of online help systems
Education	Master of Science, Instructional Design San Diego State University, 2002 Bachelor of Science in Professional Communication New Mexico Tech, 1999
Instructional Design Skills	Designed and developed Web-based training, including job aids and self-paced tutorials for Fortune 500 clients Helped more than 20 companies design training programs to support long-term organizational goals Developed presentations using Dreamweaver and PowerPoint
Writing Skills	Wrote student workbooks to support 15 different classroom training courses Authored approximately 20 different user manuals and technical guides
Evaluation and Assessment Skills	Provided formative and summative evaluations on training for more than 25 clients Designed course assessment surveys for all training programs I designed Conducted more than 100 subject matter expert interviews
Employment History	Instructional Designer, E-Learning Associates, 2002–present Technical writer, Smith and Associates, 1999–2001
Software Knowledge	Word, Dreamweaver, PowerPoint, CBT authoring tools, and e-learning development technologies
Professional Memberships	American Society for Training and Development Society for Technical Communication
Portfolio	Available on request

FIGURE 8.5
Sample Functional Resume

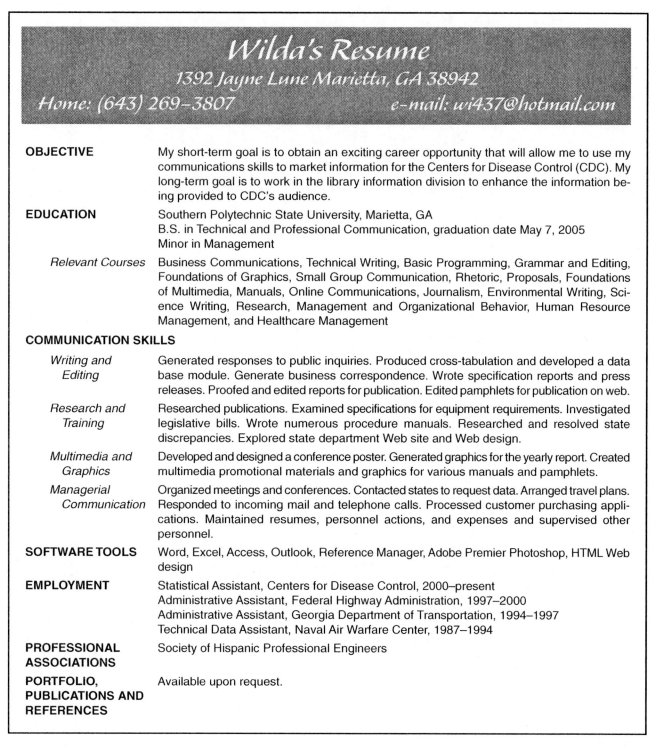

Wilda's Resume
1392 Jayne Lune Marietta, GA 38942
Home: (643) 269-3807   **e-mail: wi437@hotmail.com**

OBJECTIVE	My short-term goal is to obtain an exciting career opportunity that will allow me to use my communications skills to market information for the Centers for Disease Control (CDC). My long-term goal is to work in the library information division to enhance the information being provided to CDC's audience.
EDUCATION	Southern Polytechnic State University, Marietta, GA B.S. in Technical and Professional Communication, graduation date May 7, 2005 Minor in Management
Relevant Courses	Business Communications, Technical Writing, Basic Programming, Grammar and Editing, Foundations of Graphics, Small Group Communication, Rhetoric, Proposals, Foundations of Multimedia, Manuals, Online Communications, Journalism, Environmental Writing, Science Writing, Research, Management and Organizational Behavior, Human Resource Management, and Healthcare Management

COMMUNICATION SKILLS

Writing and Editing	Generated responses to public inquiries. Produced cross-tabulation and developed a data base module. Generate business correspondence. Wrote specification reports and press releases. Proofed and edited reports for publication. Edited pamphlets for publication on web.
Research and Training	Researched publications. Examined specifications for equipment requirements. Investigated legislative bills. Wrote numerous procedure manuals. Researched and resolved state discrepancies. Explored state department Web site and Web design.
Multimedia and Graphics	Developed and designed a conference poster. Generated graphics for the yearly report. Created multimedia promotional materials and graphics for various manuals and pamphlets.
Managerial Communication	Organized meetings and conferences. Contacted states to request data. Arranged travel plans. Responded to incoming mail and telephone calls. Processed customer purchasing applications. Maintained resumes, personnel actions, and expenses and supervised other personnel.
SOFTWARE TOOLS	Word, Excel, Access, Outlook, Reference Manager, Adobe Premier Photoshop, HTML Web design
EMPLOYMENT	Statistical Assistant, Centers for Disease Control, 2000–present Administrative Assistant, Federal Highway Administration, 1997–2000 Administrative Assistant, Georgia Department of Transportation, 1994–1997 Technical Data Assistant, Naval Air Warfare Center, 1987–1994
PROFESSIONAL ASSOCIATIONS	Society of Hispanic Professional Engineers
PORTFOLIO, PUBLICATIONS AND REFERENCES	Available upon request.

FIGURE 8.6
Sample Combination Resume
Source: Used with permission from Wilda Parker. Personal information adapted for publication.

combination resume. She has included a communication skills section with subheadings for writing and editing, research and training, multimedia and graphics, and managerial communication in order to emphasize her strongest qualifications. Wilda has deemphasized an employment record unrelated to technical communication by placing this information near the bottom of her resume. She has also mentioned her portfolio on her resume.

Scannable resume.   Employers today often receive hundreds of resumes for a single job opening. Due to the sheer volume of responses, more and more employers are requiring job applicants to submit resumes in a scannable format. You should prepare a scannable resume now in case an employer asks you for one later on. As usual, technology has provided an easy way for employers to review hundreds of resumes electronically by using an optical scanner to search for keywords and phrases that match the language in the job description. The more matches or hits your resume receives, the better your chances of being invited in for an interview. Figure 8.7 contains ten valuable tips on how to create a scannable resume.

Content of a scannable resume.   While the format for a scannable resume is very different from that of a traditional resume, the content isn't except for the keyword summary. Consider using the same categories as in your traditional resume: contact information, objective, education, employment summary, skills and software tools, certifications, languages, references, and mention of a portfolio. Also, remember to use the language of the job whenever it is appropriate to describe your skills and abilities. You may want to do this by including a background or qualifications summary section, as you would in a functional resume, immediately after your career objective.

How to create a scannable resume.   The following steps can be used to convert your traditional resume into a scannable version. Remember to follow the tips on content and format mentioned earlier:

1. Open your resume in Word.
2. Eliminate any special formatting and any graphics to make your resume scanner-friendly.
3. Use the "save as" option to save your resume and select "text only with line breaks."
4. Print your resume in this plain text format, revise where needed, and proofread it carefully several times.

The resume in Figure 8.8 is a scannable version of Frank Prater's chronological resume. Notice how white space is used liberally to create breaks between sections of the resume. Another option is to submit your

---

1. Eliminate traditional resume formatting. Use capital letters to emphasize important information. Do not use italics, underlining, horizontal or vertical lines.

2. Include a keyword summary at the top of your resume.

3. Use the language of the job ad whenever possible to increase your chances of getting hits or matches.

4. Use a standard typeface such as Times, Arial, Optima, Futura, or Universal and use a font size from 10 to 14.

5. Avoid columns. Scanners move from left to right.

6. Avoid boxes and graphics

7. Keep the maximum number of characters per line to 65.

8. Use black ink on white or a very light color 8 1/2" × 11" paper.

9. Use only a laser-quality printer.

10. Remember to use plenty of white space even if it means your scannable resume is on more than one page. Place your name on each page.

FIGURE 8.7
Ten Tips for Formatting a Scannable Resume
Source: U.S. Department of Labor, Office of Disability Employment, 2005 (http://www.dol.gov/odep/pubs).

FRANK PRATER

1390 Trailedge Way
Malone, NY 16789
Home: 315.723.0391
Cell: 315.534.8932
fprater@earthlink.net

KEYWORDS: technical writing, document design, multimedia, computer science, database structures, Java, programming languages

OBJECTIVE

Technical writing position using my writing skills and strong technical background

QUALIFICATIONS SUMMARY

A diverse communications background focusing on document design, multimedia design, and different technical writing projects supplemented by knowledge of database structures and programming languages

EDUCATION

B.S., Technical and Professional Communication
Minor: Computer Science
Malone University, Malone, NY

RELEVANT COURSES

Advanced Grammar and Editing
Journalism, User Documentation, and Proposal Writing
Public Speaking, Science Writing, and Technical Training
C++ Programming and Database Systems
Java Programming

WORK EXPERIENCE

Database Developer, Malone Sweeteners, June 2002–present

- Designed and implemented a database to track customers and products
- Completed data entry of existing customers, contacts, and products
- Wrote a user's manual for the database

Cashier/Lot Assistant, Canton Nurseries, 2000–2001

Assistant Manager, Super Video, 1999–2000

SOFTWARE TOOLS

- Microsoft Office Suite
- Adobe FrameMaker
- Adobe PageMaker
- Adobe Photoshop
- HTML

ACHIEVEMENTS

- Outstanding Technical and Professional Communications Senior, 2003
- 2nd Degree Taekwondo Black Belt

PORTFOLIOS AND REFERENCES

- Professional portfolio available at interview
- Electronic portfolio available upon request
- References available upon request

FIGURE 8.8
Sample Scannable Resume

resume to an employer as e-mail text or to send it as an attachment to an e-mail letter of application. Since many people today will not open attachments from individuals they don't know, you may decide to include your text-formatted resume in the body of your e-mail. Include reference numbers of job ads on the subject line of your e-mail. If you have a Web-hosted resume, you may want to include that URL in your e-mail so that the employer sees a fully formatted professional resume.

Writing the letter of application/cover letter

Even if you haven't written a letter of application, also called a "cover letter," for a full-time position, you have probably written a letter of application for another reason, such as an internship, summer job, or part-time position. You should write a letter of application for each resume you mail. The purpose of the letter of application is twofold: First, you want to convince the employer that you are the best person to fill the position; second, you want to ask for an interview.

Tips for writing an effective letter of application

While a resume is essentially a fact sheet that summarizes your qualifications, a letter of application is a narrative that markets your candidacy by showing the employer how your qualifications match the job description. A well-written letter also gives you a chance to show your personality and highlights, or expands upon, particular accomplishments. It should be positive and persuasive. Figure 8.9 presents 10 tips on how to write an effective letter of application.

Organizing the letter of application

Most successful application letters have a three-part structure: a short opening paragraph; two to four body paragraphs providing evidence that you can do (or have done) what you claim, noting the benefits for the company; and a closing paragraph asking for the interview.

---

1. Personalize each letter to fit the advertised needs of each company's job description. Avoid generic "one size fits all" letters of application.

2. Know the job ad and respond specifically to what is in it.

3. Address your letter to a specific person in the company, if possible, rather than a department (i.e., Human Resources). Make sure that you spell that person's name correctly.

4. Focus on the company—the "you" in your letter. Avoid using the first-person pronoun (e.g., "I") excessively in your letter. Doing so gives the appearance of boasting.

5. Focus on how your skill sets match what the company is looking for.

6. Keep your letter to one page but support your key qualifications with specific details.

7. Refer to your resume and portfolio in your letter.

8. Use high-quality stationery and envelopes.

9. Ask for the interview. Don't be timid.

10. Edit and proofread several times. Like your resume, your letter of application is a writing sample.

FIGURE 8.9
Ten Tips for Writing a Successful Letter of Application

Tips for writing the opening paragraph. In the opening paragraph, state that you are applying for a position and refer to that position specifically by name. The letter should not seem to be an inquiry about a job or a request for information. In the opening paragraph, also indicate what resource or resources you used to find the position (e.g., a professional journal, Web site, career center, head hunter, newspaper, teacher, or work associate). If the employer has used a job identification number, include it. You may want to write your opening paragraph using one of these techniques:

* Drop a name.
* Use a summary.
* Stress your strongest qualification.

Examples of opening paragraphs using these techniques. If you decide to use a contact's name in your application letter, let that person know that you are doing so. More likely than not, the employer will call or write to your contact for information about you.

Dropping a Name

Dr. James Smith, Professor of Professional Communication and Chair of the same department at Midwestern University, has informed me that Bancroft and Associates has a full-time position for a multimedia specialist. Please consider me an applicant for that position.

Using a Summary

Four years of courses in professional communication at Georgia Polytechnic, plus two summers interning at the Centers for Disease Control, have given me the knowledge and work experience needed for the entry-level writing position that you describe in the June issue of the *AMWA Journal.* Please consider me an applicant for that writing position.

Stressing Your Strongest Qualification

My knowledge of RoboHelp and FrameMaker, two of the required software packages mentioned in your ad, will help me contribute immediately to your software documentation projects. Please consider me an applicant for the user documentation position.

Tips on writing the body paragraphs. Make it clear why you are interested in the company and the position. Make sure that the body paragraphs show (not merely claim) your strongest selling point, whether it is your education, work experience, or both. If you have limited related work experience, explain how your education makes you a qualified applicant. If you have relevant work experience, point out how key projects and activities that you worked on provide the skills that will transfer well to this new position. Demonstrate how you can help the company and not how the company can help you. Tie your letter to your resume but avoid merely repeating the information there. Refer specifically to your portfolio, noting how the skills displayed there clearly match the job requirements.

Example

My portfolio shows my understanding of various designs for electronic and paper-based documents. Several of the graphic tools that you require are the same ones that I have used in my portfolio.

If you don't have much related work experience, don't be negative. Instead, try to compensate by focusing on more general character traits such as a strong work ethic.

Example

You may consider it important that I earned more than 80 percent of my college expenses doing sales work during summer vacations. As my resume shows, I have worked as a successful telemarketer for three large Chicago-based companies. I have also done volunteer work for the March of Dimes, designing a brochure and several marketing pieces for fund-raisers.

Tips on writing the closing paragraph.   The closing paragraph of an application letter has one function: to ask for an interview. Indicate your flexibility in regard to time, and make it easy for the employer to contact you, providing telephone numbers and e-mail addresses where you can be reached. Do not imply that you want an interview; ask directly for one. Figures 8.10 and 8.11 are sample application letters.

What not to say in your letter of application/cover letter

It's just as important to remember what not to say in your letter of application as what to say. Figure 8.12 provides nine tips on what to avoid.

July 18, 2004

Ms. Tracey Haddack
Marketing Manager
Just-n-Time Marketing and Customer Services
Suite 234
143 Piedmont Place
Macon, GA 33008

Dear Ms. Haddack:

Please consider me a candidate for the PowerPoint presentation designer position (ID# 37942) that you advertised in *The Marketing News* database on July 16, 2004. The PowerPoint presentations I have created, in addition to my degree in international technical communication, match up well with the requirements you mention in your ad.

While pursuing my degree at Southern Polytechnic State University, I completed courses in multimedia, graphics, technical writing, business communication, and international technical communication. Projects for these classes allowed me to develop my graphic and multimedia skills using such tools as Microsoft PowerPoint, Adobe Photoshop, Adobe Illustrator, and Adobe PageMaker, software applications that you mention in your ad.

You may be interested to learn that I am currently working on a project to design a training session to instruct graduate students on how to use Microsoft Producer, a stream media add-on for PowerPoint. Once complete, this project will be included in my portfolio, which contains two other professional-quality PowerPoint presentations. My electronic portfolio is available upon request.

May I have an interview at your earliest convenience to further discuss my qualifications listed on the enclosed resume? You can reach me anytime by calling my cell phone (770.946.9154) or by e-mail (ngonzalez@hotmail.com) so that we can arrange a convenient time for an interview.

Sincerely,

Norma O. Gonzalez

Enc: resume

FIGURE 8.10
Sample Application Letter
Source: Norma O. Gonzalez. Personal information adapted for publication.

Mr. Jason Dombrowsky
Human Resources Manager
Simi Valley Software Systems
1490 Sunset Avenue
Simi Valley, CA 93062

Dear Mr. Dombrowsky:

Your advertisement for a junior technical writer in the Careerbuilder.com database dated July 16, 2004, fits perfectly with my qualifications and career goals. Please consider me a candidate for that position.

Your job requirements mention experience in writing software documentation. My consulting experience writing software documentation for Janes and Associates Software should match up well with that requirement. I have also worked as a professional reporter and as the editor of two college newspapers. As a consultant, journalist, and editor, I have demonstrated that I can effectively facilitate meetings with users, subject matter experts, and staff members to gather appropriate information. All of these work experiences will translate well into facilitating meetings with multiple audiences to derive documentation requirements. As you can see on my resume, I have all the appropriate qualifications to meet Simi Valley Software Systems' needs, including the correct software knowledge. Furthermore, I work well under minimum direct supervision and have proven abilities in both document creation and editing, as shown in my electronic portfolio, which is available upon request.

May I have an interview at your convenience so we can further discuss my qualifications for this position? You can reach me at my cell phone (937.726.8324) anytime or at eagertobegin@comcast.net. I look forward to hearing from you.

Sincerely,

Miranda Bennett
Cell: 937.726.8324
eagertobegin@comcast.net

Enc: resume

FIGURE 8.11
Another Sample Application Letter
Source: Used with permission, Miranda Bennett 2005. Personal information adapted for publication.

1. Don't write an autobiography. Try to eliminate excessive use of the personal pronoun "I."
2. Avoid using a negative tone.
3. Don't boast; instead, focus on how you can help the company.
4. Don't get too personal.
5. Don't be long-winded.
6. Eliminate all grammar and punctuation errors.
7. Don't misspell the employer's name. Don't misspell anything.
8. Avoid mentioning salary requirements.
9. Don't forget to mention your resume and portfolio in your letter.

FIGURE 8.12
What to Avoid in Your Letter of Application

SUMMARY

Conducting a successful job search is hard work. Chapter 8 discusses how you can make this process easier by developing effective job-search strategies that help you match your career goals with special fields within technical and professional communication. The chapter also provides tips on how to develop a marketing plan, how to network, and how to write an effective resume and letter of application.

ASSIGNMENTS

Assignment 1: *Finding Career Information on the Internet*

Using the Internet, locate five Web sites that provide career information for professional communicators. That information could include job postings, resume services, or any of the career information mentioned in Chapter 8. Write down the URL and a short description of each site's content. Comment on the value of the information and prepare to share it with classmates or your mentor.

Assignment 2: *Creating/Updating Your Resume*

Update your resume or create a new one using one or more of the resume styles: (chronological, functional, combination) discussed in this chapter. To better market your qualifications, you will probably want to use more than one style, as well as a scannable version. For this assignment, your resume or resumes should focus on a specific career within professional communication (e.g., Web design, corporate communications, graphic design). In groups of three, exchange resumes and offer each other oral and written feedback.

Assignment 3: *Preparing a Scannable Resume*

Prepare a scannable version of the resume you created in Assignment 2.

Assignment 4: *Writing a Sample Letter of Application*

Write a letter of application in response to a job description. You might want to start by searching databases at your college's career center, by responding to a print ad in a professional journal, or by using the Internet. Attach the resumes that you created in Assignments 2 and 3.

REFERENCES

Bloch, Janel M. "Online Job Searching: Clicking Your Way to Employment." *Intercom* 50, no. 8 (September– October 2003): 11–14.

Bolles, Richard. *Job Hunting on the Internet*, 2nd ed. Berkeley, CA: Ten Speed Press, 1999.

———. *What Color Is Your Parachute? A Practical Manual for Job-Hunters and Career-Changers*, 33rd ed. Berkeley, CA: Ten Speed Press, 2004.

Conlin, Michelle. "And Now, The Just-In-Time Employee." *Business Week* 28 (August 21–28, 2000): 169–170. http://web27.epnet.com/citation.asp?tb (accessed July 28, 2005).

Crowther, Karmen N.T. "How to Research Companies." In *Planning Job Choices 2000*, 43rd ed. Bethlehem, PA: National Association of Colleges and Employers, 2000.

Lannon, John M. *Technical Communication*, 9th ed. New York: Longman, 2003.

Pfeiffer, William Sanborn. *Technical Writing: A Practical Approach*, 5th ed. Upper Saddle River, NJ: Prentice Hall, 2003.

U.S. Bureau of Labor. "Writers and Editors Job Outlook." *U.S. Bureau of Labor Statistics Occupational Outlook Handbook.* 2004–05 ed. http://www.bls.gov/oco/ocos089.htm (accessed April 22, 2005).

U.S. Department of Labor, Employment and Training Administration. "Tips for Finding the Right Job" (1996). http://www.doleta.gov/uses/tip4jobs.pdf (accessed May 25, 2004).

U.S. Department of Labor, Office of Disability Employment Policy. "Writing and Formatting a Scannable Resume." (February 5, 2005): 1–2. http://www.dol.gov/odep/pubs/ek99/resume.htm (accessed September 7, 2005).

Van Wicklen, Janet. *The Tech Writer's Survival Guide.* New York: Checkmark Books, 2001.

APPENDIX A: CAREER RESOURCES, WEBSITES AND URLs

American Marketing Association
(http://www.marketingpower.com)

American Medical Writers Association
(http://www.amwa.org)

American Society for Training and Development
(http://www/astd.org)

American Society of Journalists and Authors
(http://www.asja.org)

America's Job Bank
(http://www.ajb.org)

Association of Health Care Journalists
(www.healthjournalism.org)

Avue Central
(http://www.avuecentral.com)

Careerbuilder.com
(http://careerbuilder.com)

Careers at Yahoo!
(http://careers.yahoo.com)

Computer Jobs.com
(http://www.technicalwriter.computerjobs.com)

The Careers Organization
(www.careers.org)

ELance
(www.elance.com)

FlipDog.com
(http://flipdog.monster.com)

Graphic Artists Guild
(http://www.gag.org)

Hotjobs.com
(http://hotjobs.yahoo.com)

International Association of Business Communicators
(http://www.iabc.com)

International Public Relations Association
(http://www.ipra.org)

International Webmasters Association
(http://www.iwanet.org)

Job Bank USA
(http://jobbankusa.com)

Monster TRAK
(www.monstertrak.monster.com)

Monster.com
(www.monster.com)

National Association of Science Writers
(http://nasw.org)

Newbie TechWriter
(http://www.cloudnet.com/~pdunham)

Society for Technical Communication
(http://www.stc.org)

Society of Environmental Journalists
(http://www.sej.org)

Society of Publication Designers
(http://www.spd.org)

Technicalwriterjobs.com
(http://www.technicalwriterjobs.com)

TechWriting at About.com
(http://techwriting.about.com)

Techwritingjobs.com
(http://www.techwritingjobs.com)

The Usability Professionals' Association (UPA)
(http://www.upassoc.org)

Webdeveloper.com
(http://www.webdeveloper.com)

The Write Jobs
(http://www.writejobs.com/jobs)

Writerfind.com
(http://www.writerfind.com)

9 Using Portfolios During Interviews

Portfolios are great for job interviews. They give the applicant more control over the interview. They also provide support and talking points when you answer questions about skills and experience. Tom

INTRODUCTION

As Tom notes, your portfolio can help showcase your skills and experience; it can also illustrate what you claim on your resume. As noted in Chapter 8, an application letter and a resume are designed to achieve one goal: to get you an interview with the hiring manager, human relations manager, documentation manager, or other company representative whose job is to select the best person for the advertised position. Interviews can last for 1 hour or less, several hours, or even several days. Regardless of the length of time, the goal of the interview is the same—to hire the candidate who best matches the qualifications outlined in the job description and who is the best fit with the company. You, of course, want to be that candidate, and using your portfolio to your advantage will help you get that job offer.

Chapter 9 focuses exclusively on the job interview—how to prepare for it and how to use your portfolio during the interviewing process. Preparing for the job interview requires you to research thoroughly each company you plan to interview with so that you are familiar with its products, services, management philosophy, and, when possible, its approach to interviewing. Because your paper and electronic portfolios are your most important interview resources, you should know their contents thoroughly and feel comfortable talking about your projects. Chapter 9 provides guidelines so that you will be ready when you start interviewing.

Volumes have been written on job interviews, and on the Internet you will find thousands of sites that can help you prepare. Specifically, Chapter 9 focuses on the following topics:

* Preparing for the job interview
* Using your portfolio during the job interview
* Identifying and handling illegal questions
* Discussing salary
* Responding to the job interview

PREPARING FOR THE JOB INTERVIEW

Studies often show that the number one reason job applicants do poorly during an interview is that they know very little about the company offering the job. Table 8.2 in Chapter 8 identified some key resources that provide specific company information, and some of that advice deserves repetition here. Visit the company's Web site to see how the company markets itself; review the company's management philosophy and plans for growth, and note any new products or software releases. To get a more objective viewpoint, do an Internet keyword search using the company's name. You may uncover an article or two that comments on the company's financial health and possible mergers that could mean downsizing. Network within professional associations like the Society for Technical Communication, the International Webmasters Association, the American Society of Journalists and Authors, and the American Society for Training and Development, for example, to find out what other professionals say about the company. Once you have done your homework and researched the company, you can formulate specific questions to ask at the interview, and you can make an initial decision as to whether or not you will be comfortable working for this company.

Sometimes the published job description will differ from what the hiring manager is really looking for (Molisani 2003, 21). For example, the published job description may focus on the requirements for a software documentation writer who will write external user manuals and standard operating procedures (SOPs). The hiring manager may also be looking for someone who can do some in-house training as it relates to the SOPs, even though this need may not be addressed in the published job description. You might find a fuller description of the job position by checking the company's Web site, in this case to see what kinds of training activities, if any, the company engages in. If the posted job position provides a contact for further information and notes that a complete job description is available on request or on the company's Web site, you may want to do a follow-up. Another way to get a full job description is to ask the interviewer for one early in the interview. The more you know about the job, the better you will be able to tailor your responses to the company's true needs.

Know what to expect

Try to find out as much as you can about the company's interviewing process. While each company conducts interviews differently, companies have similar objectives in terms of what they hope to accomplish during the interview. The screening, or initial, interview is designed to identify a short list of job applicants to invite for a more extensive behavior-based interview. The screening interview is normally designed to last about 30 minutes. One large Fortune 500 company trains its interviewers to structure this interview as follows:

* Introduction (1 to 2 minutes to establish rapport and to put the candidate at ease)
* Information-getting segment (roughly 12 minutes to get information about the candidate through a series of open-ended and probing questions)
* Information-giving segment (roughly 12 minutes to present company information such as the work environment, related job responsibilities, training, relocation, and advancement opportunities as they apply to the applicant's job interests)
* Closing (1-minute summary, also noting follow-up actions)
* Record keeping (3 minutes to record specific observations about the candidate's qualifications and character traits)

Whatever the organization of the interview, remember the 50/50 rule: You will probably speak half of the time and listen half of the time (Bolles 2004, 243). Your portfolio can help you make the most of your speaking time.

Prepare answers to questions you might be asked

You will find a wide range of books, articles, and Web sites that provide lists of sample questions often asked at screening interviews. You should write out answers to the questions you think an interviewer is likely to ask. Do not bring the written answers to the interview. Published lists of questions often include those listed in Figure 9.1.

1. Why should we hire you?

2. What are your major strengths and weaknesses?

3. Tell me a little about yourself.

4. What do you see yourself doing 5 years from now?

5. Why did you leave your last job?

6. Can you explain this gap in your employment history?

7. What motivates you to do your best work?

8. Do you work well as part of a team? Please provide an example.

9. Do you think of yourself as a leader? Please provide an example.

10. Why did you choose your major?

11. What college subjects did you like best and least? Why?

12. What two or three accomplishments have given you the most satisfaction? Why?

13. What was the best decision you ever made?

14. In what kind of work environment are you most comfortable?

15. How do you work under pressure? Can you give me an example?

16. Are you willing to relocate?

17. What do you know about our company?

18. What contributions can you make to our company?

19. What have you learned from your mistakes?

20. Have you ever had a problem with a supervisor? How did you resolve it?

FIGURE 9.1
Frequently Asked Questions at Screening Interviews

While it may be difficult to prepare answers to all 20 of these questions, you should develop answers to the 4 core questions that many interviewers ask:

1. Tell me about yourself.
2. Why should I hire you?
3. How well do you work with deadlines?
4. Why are you leaving your current employer?

When answering these questions, keep in mind the 2-minute rule; try to respond with an answer that is roughly 2 minutes long. Two minutes allows you to develop your response in detail without rambling and, perhaps, revealing a weakness that would be best left uncovered.

EXERCISE 9.1 PREPARING FOR AN INTERVIEW

Write down answers to the four core questions listed above. Keep your answers focused, and practice them so that each answer is roughly 2 minutes long.

Inventory your soft skills

If you are applying for an entry-level technical communications position and have little or no related work experience, you should inventory your soft skills and use them to bolster either your limited technical skills or related work experience. Soft skills are those that can't be easily measured or quantified. You may want to refer back to the work you completed in Chapter 2 on your professional identity. Very often companies look for the following soft skills when interviewing candidates for entry-level technical communication positions:

* Research and interview skills
* Time management skills
* Ability to work well in teams
* Willingness to learn
* Creative thinking and problem-solving skills
* Ability to multitask
* Openness to work with all personality types

Since the field of technical and professional communication is continually changing, these core soft skills will be helpful during job interviews if you can use them to demonstrate how they will help you perform the job. Your portfolio will help you show projects that require these skills. You may also want to reflect back on the internships, part-time and full-time positions, volunteer work, and committee work that you have done and isolate each soft skill you have listed. Pursuing your degree, raising a family, and working part-time or full-time are good examples of your multitasking ability.

Prepare for behavior-based interview questions

Some interviewers ask behavior-based questions during the screening interview but, more often than not, you will be asked these types of questions during the second or third interview that often is held onsite. Behavior-based interview questions permit the interviewer to evaluate how you have reacted or behaved in certain types of situations. Roughly 150,000 managers are converting to this style of interviewing each year, and some companies, like Hewlett-Packard and AT&T, train all of their managers in behavior-based interviewing strategies (Washington 2004, 107). Companies that regularly hire technical writers, for example, often use two types of interviews: the screening interview and the onsite behavior-based interview. For example, a telecommunications company might employ 15 technical writers whose main responsibility is to create and maintain documentation for several Internet products. These writers may support three to seven product releases a year, amounting to 30K pages and 160 books. The documentation may be distributed to customers via CD, the Web, and limited paper formats. It is not unusual for an onsite interview to last 2 hours or longer and involve two or more interviewers. During the interview, you may be evaluated according to criteria similar to these:

* Technical writing and experience
* Interpersonal effectiveness and communication skills
* Software knowledge
* Project management skills
* Customer orientation
* Problem-solving and thinking skills

The interview might include 15 or more behavior-based questions. Sample behavior-based questions are the following:

* Describe a situation where a subject matter expert you needed to talk to didn't return your call. What did you do next?
* Describe a situation where you had to handle conflict.
* Tell me about a situation where you missed a deadline. How did you handle it?

How you respond to such questions will help the interviewer determine not only key behavioral characteristics that might be important on the job but also how well you react to tough questions under pressure. Using your portfolio whenever possible to show projects that address the kinds of behavior the interviewer is interested in can work to your advantage. You will probably have one or more group projects in your portfolio that were successfully completed only after group conflicts were resolved. Now is your chance to discuss them. How you answer such questions as "Describe your conditions for best work" would provide the interviewer with important feedback on the type of work environment you prefer.

Other examples of behavior-based questions are as follows:

* Provide a specific example of a time when a co-worker or classmate criticized your work in front of others. How did you react? How did this criticism influence the way you communicate with others?
* Describe the approach you use for tracking multiple projects. How does this approach help you meet deadlines?
* Discuss a time when you were able to get co-workers or classmates to work together. How did you do this? What was the result?

  Source: "How to Behave in a Behavior-Based Interview" 2000, 64.

The popularity of the behavior-based interview has resulted in a number of publications that you can easily find on the Internet or in your local bookstore.

EXERCISE 9.2 RESPONDING TO BEHAVIOR-BASED QUESTIONS

Write down responses to the sample behavior-based questions listed here. Remember to connect your responses to projects in your portfolio.

While you can't prepare for every question that you might be asked, try to anticipate the odd question that seems to come out of nowhere. Most interviewers want to see if you can think on your feet. Instead of asking this question directly, allowing job applicants to respond with a specific example, one interviewer likes to ask them to explain their philosophy of life, forcing them to think quickly (Kaffer 2002). While you probably won't be asked this question, listen carefully to what the interviewer says and be ready to respond. Reviewing the work that you did in developing your professional identity may help you prepare for this kind of question.

Develop your own list of questions

When an interviewer asks if you have any questions, have two or three prepared. If you have no questions, an interviewer may interpret your silence as apathy, weak communication skills, or lack of company research. Avoid questions regarding salary (if it is unknown), and let the interviewer bring up this subject. If this doesn't happen by the end of the interview, feel free to raise the topic. Always ask questions about advancement opportunities and performance reviews. During the interview, you will have opportunities to ask for clarification and follow-up questions in response to information that the interviewer is providing. You should also have some open-ended questions to ask during the interview such as these:

* What types of projects would I be working on?
* What would I be expected to accomplish during my first 6 months on the job?
* When can I expect to undergo my first performance review?
* What would be my primary responsibilities?
* What is the organizational and reporting structure?
* When may I expect to hear from you?

Practice, practice, practice

You can improve your chances of succeeding at the interview by rehearsing the situation and by visualizing a positive outcome. The mock, or role-playing, interview is one of the best techniques to use. Often university career centers provide this service; if yours does, take full advantage of it. If not, ask your mentor or a friend to work with you on a practice interview. Provide a list of questions that you expect to be asked at the interview and have this person choose random questions from the list. To create an organizational context for the interview, give your mentor or friend information about the company and the position. Thoroughly practice using your portfolio during these mock interviews until you become comfortable discussing the documents in your portfolio and can turn effortlessly to any document you need to respond fully to a question. If possible, have the mock interview videotaped or at least audiotaped. When you review your performance, note the questions that may have posed problems. If your interview was videotaped, look for nonverbal problems that may detract from your performance, such as poor eye contact or lack of enthusiasm in responding to questions.

Plan what to wear

A great deal has been written about what to wear to an interview, and an Internet search will provide useful information. Learn as much as you can about the company's organizational culture before the interview so that you can dress to fit in. Most publications on interviewing warn against dressing in a trendy or casual way even if the company's dress code is business casual. The emphasis is on appropriate attire; if you are in doubt, wear a suit. It's normally better to be dressed too formally rather than too casually.

In particular, if you are applying for a communication position in public relations, training, or some other role where you may be representing the company in a very visible way, dress conservatively. A safe choice for both men and women is a clean, freshly pressed business suit. Try to opt for conservative colors such as navy or medium to dark gray. Wool or wool blends are good fabrics for all seasons. If your onsite interview requires that you stay for more than 1 day, make sure that you pack more than one suit. Safe colors for men's shirts include white and conservative solid colors like blue or gray. Hair should be neat and clean; facial hair should be well groomed. Fingernails should be clean, and press-on nails should not be too long. Cologne or perfume should be used sparingly, and shoes should be shined. If you are wearing heels, make sure you can walk comfortably in them in case the onsite interview includes a tour of the company's facilities.

USING YOUR PORTFOLIO DURING THE JOB INTERVIEW

This section provides tips on how to use your portfolio effectively during the interview.

Integrate your portfolio into the interview

Sarah, a recent graduate of a technical and professional communication program, offers the following advice on how to use your portfolio in response to questions that you might be asked at an interview:

> Once a potential employer has granted you an interview, bring your portfolio with you. I found that employers would not ask to see it, but I would generally bring it into a conversation by showing my work as answers to some of the questions I was asked. Granted, it's relatively awkward at first, but once you get used to showing off your work, it becomes very natural. Also, show the portfolio to anyone who will sit down and look at it. Doing this will often get you reference and job opportunities that you might never have known about.

You will find many opportunities to work your portfolio into the interview. If the interviewer doesn't ask to see it, use your portfolio to respond to such questions as "Tell me about yourself" or "Why should I hire you?" by showing your professional samples. You can always use the direct approach and ask the interviewer if he or she would like to see samples of your work. What interviewer can say "no" to that question?

Tailor your portfolio for each interview

Organize your portfolio to meet the requirements of the position for which you are applying. After presenting your resume, try to open with your most relevant piece. If you are applying for a position that involves designing multimedia presentations, screen captures of key slides may be helpful early in your paper portfolio, along with a storyboard or handout that shows the sequencing of slides. If you will be writing documentation, you may want to show a documentation plan, a table of contents, and sample pages from a manual you have written. Michel, a communications specialist with the CDC, described how she organized her portfolio for her interview:

> One of my showcase pieces happened to be on a topic that the lead interviewer was an expert in. He knew the subject very thoroughly and thought my work was balanced and well researched. That was a lucky break!

Indicate before the interview that you plan to bring your portfolio and ask in what medium the interviewer(s) prefer to see it.

Today, many interviewers still prefer a paper portfolio because it is easy to access and flip through. In general, paper portfolios are most effective during the interview because they allow you to structure the interview and give you a chance to discuss the included projects that best represent your qualifications for the job. Electronic portfolios are dynamic and interactive and may soon edge out the paper portfolio as the preferred format for interviews. However, it is a good idea to bring both the paper and electronic portfolios to the interview. If you are applying for a position as a Web designer or multimedia specialist, your electronic portfolio should be showcased. Before the interview, you may also want to find out what resources are available and whether you will be interviewed by one person or a panel. If you are to be interviewed by a panel, you may want to request an Internet connection and a LCD projector so that you can show the electronic portfolio. Tom notes that at an interview he was asked to present his electronic portfolio to a technical documentation department and chief engineer. He comments:

> Fortunately, I was able to project my online portfolio onto a digital screen and pass my printed portfolio around to the gathered department. I don't know how I would have handled the situation without these two assets.

Showcasing your portfolios immediately displays the tools you are familiar with and allows you to control the presentation by focusing on the projects that align most closely with the job requirements. It is a good idea to have several copies of sample pages from the projects you want to emphasize to be used as handouts. Use familiar formats such as pdf., doc., ppt., and html so that the interviewer will feel comfortable downloading your files. Interviewers often like thumbnail links to documents, so you may want to use them as well.

Do not include anything that someone can criticize

Your portfolio should contain your best work, and it should be of professional quality. Including anything else would be a mistake.

Know your portfolio backward and forward

Your portfolio will be used at the interview to prove what you claim on your resume and to demonstrate and elaborate on your strengths as you specifically name them. To do this effectively, you must be able to navigate the portfolio flawlessly to find the samples that respond best to the interviewer's question. You don't want to waste time flipping through the paper portfolio or clicking on links in the electronic portfolio in a desperate search for the sample you need. If the interviewer wants to see an example of a project done in PageMaker™, you should be able to turn to it or click on it right away. Tabbing or color coding your paper portfolio will allow you to reference key projects immediately. If there is an introduction to each project noting what it is and what skills the piece emphasizes, you will be able to comment on the piece with little trouble. If you are asked for a sample project created in RoboHelp™ or Dreamweaver™, it may be helpful to have a CD available, perhaps in a cellophane pocket, with sample help files or Web pages for viewing during the interview or to leave behind afterward.

Discuss any team efforts

Most interviewers want to see evidence of your ability to work productively on a team. To help you discuss your team project at an interview and your role in creating it, you may want to include a project plan. This plan might be less than a page, with a brief project description, team member responsibilities, a time line (perhaps a Gantt chart) showing key project deadlines, obstacles encountered, and a short assessment of the project's merits. To demonstrate team skills, Sarah included in her electronic portfolio a PowerPoint presentation on South Korean business practices. Her paper portfolio included sample PowerPoint slides and an outline of her presentation. Each member of the four-member team researched, designed, and then presented on a different area of this subject, including South Korean business etiquette, cuisine, and dining habits, meetings and strategies, and South Korean culture in general.

Sarah's part in the presentation was to deliver the introduction and conclusion and to serve as moderator. In preparation for an interview, Sarah may want to include a brief project plan, as described here, because she knows that she might be nervous at the interview. The project plan would give her a focal point in discussing how the project evolved and provide an opportunity for Sarah to discuss the collaborative efforts of each team member, focusing specifically on her own contributions to the project.

In her portfolio, Nanette included a different type of group project. Hers was a Web site redesign for a nonprofit baseball league for children with disabilities. Like Sarah, Nanette may want to include a brief project plan that will help her feel more at ease when she discusses the project and her contribution to it during the interview.

Plan to show before-and-after samples

Showing before-and-after samples can clearly demonstrate your problem-solving skills (Molisani 2003, 22). Even if the interviewer doesn't ask to see this type of document, showing it demonstrates how you took a poorly designed document with no graphics or misused graphics and transformed it into a well-designed professional document. Some examples of before-and-after samples are an original document marked with copyediting symbols followed by an edited revision, redesigned Web sites, or a document plan accompanied by the finished document.

Leave behind copies of sample projects that are requested

If the interviewer requests that you leave behind any sample projects, be prepared to do so. Never leave behind your original portfolio or original samples. The easiest way to handle a request for copies is to leave behind a CD version of your portfolio. The CD business card is an impressive delivery method for samples. You may also want to leave behind a packet containing copies of representative projects from your paper portfolio.

MORE INTERVIEWING TIPS

- Dress appropriately.
- Turn off your cell phone.
- Arrive 15 minutes before your interview.
- Visit the restroom before your interview and check your appearance in the mirror.
- Remember the interviewer's name, and ask for a business card before you leave.
- Never criticize a previous employer. Your interviewer may assume a negative attitude and conclude that you will say similar things about his or her company.
- Don't appear desperate.
- Bring a PDA or pen and paper to take notes.

Table 9.1 provides a brief checklist that will help you prepare for your interviews.

TABLE 9.1
Interview Preparation Checklist

✓ Before the Interview	✓ During the Interview	✓ After the Interview
Research each company that you plan to interview with carefully.	Arrive at least 15 minutes early.	Send a short thank-you letter to the person or people who interviewed you.
Get a full description of the position.	Listen attentively for questions that will allow you to show your portfolio during the interview (e.g., "Tell me about yourself," "Why should we hire you?").	Follow up with any additional information that was requested.
Find out about the company's interviewing process, if possible.	Talk about your showcase pieces that best match your skills with the job description.	Keep thorough records including names, dates, hiring deadlines, and offers received.
Develop a list of questions you might be asked.	Maintain eye contact with the interviewer and follow the interviewer's lead.	
Prepare answers to those questions.	Ask for the interviewer's business card.	
Develop a list of questions to ask at the interview.	Never criticize a former employer.	
Inventory your soft skills.	Discuss a before-and-after sample.	
Practice by doing mock interviews.	Be prepared to discuss salary and benefits.	
Plan what to wear.	Leave behind samples of your best work or a CD of your portfolio.	

IDENTIFYING AND HANDLING ILLEGAL OR INAPPROPRIATE QUESTIONS

There is a wide range of sources both on the Internet and in hard copy that will help you identify and handle illegal questions asked during an interview. Here are some general points to keep in mind. Federal discrimination laws enforced by the Equal Employment Opportunity Commission prohibit interviewers from asking applicants certain questions. Questions about race, gender, ethnic background, age, religion, marital status, sexual preference, physical or mental condition, or other discriminatory factors are an invasion of privacy and have no bearing on your ability to do a job. These questions do not pertain to bona fide occupational qualifications.

Almost all interviewers are familiar with the legal aspects of interviewing, but you might run into an interviewer who isn't. First, try to determine if there is an underlying motive for the question or if the interviewer has asked it out of ignorance. An interviewer may ask a personal question out of ignorance as an ice breaker at the beginning of an interview. If an interviewer asks "Are you married?" or "Do you have children?" (illegal questions), you may respond directly if you are not uncomfortable doing so. If you feel that the interviewer has no hidden agenda, you may want to respond to the underlying issue—the interviewer wants to find out if you can travel and/or work on weekends, to which your response might be "yes." You can sidestep the question by refocusing on what you hope is the unstated motive for the question. For example, if an interviewer asks about your country of origin (an illegal question), you may respond by saying that you can work legally in the United States. If you feel that the interviewer has a hidden agenda, respond by saying that the personal questions being asked have no relevance to the job requirements. Finally, of course, you can refuse to answer the question, particularly if there is a pattern of bias in the types of questions being asked. If that is the case, you may want to take legal action. Table 9.2 presents some examples of illegal and legal questions. If you do assert your constitutional rights, the tone of the interview may change, and you may lose any serious chance to be offered the position. However, you probably do not want to work for a company that is willing to ignore your legal rights.

DISCUSSING SALARY

Research salary ranges in the technical and professional field before you start interviewing. Salary information is available from your college or university career center, from the U.S. Department of Labor Statistics, from Internet job-search sites, from networking contacts, and from professional organizations. The Society for Technical Communication, for example, publishes a yearly salary survey available to its membership on its Web site. The mean salary for technical communicators in the United States for 2005 was $67,520. The mean salary for those with less than 2 years' experience was $42,910; for those with a bachelor of science degree, it was $65,460 (Society for Technical Communication 2005, 5).

Consider the entire compensation package. In addition to your salary, factor in tuition reimbursement, health coverage, investment opportunities, and the pension plan. These benefits can add thousands of dollars to your salary. You should also think long-term. What opportunities are there for promotion? What is the performance review schedule? Can you be reviewed early during your first year (i.e., 3 months after starting) to demonstrate that you have met or exceeded job expectations? You will also want to determine if you are a good fit in other areas. Does the company offer flex time? Are project deadlines reasonable? What is the turnover rate for writers? Are subject matter experts cooperative? Finally, you should estimate your market

TABLE 9.2
Sample Illegal and Legal Questions

Illegal Question	Legal Question
Are you a U.S. citizen?	Are you legally authorized to work in the United States?
Are you married?	Would you be willing to relocate if necessary?
How severe is your disability?	Are you capable of carrying out the job requirements?
What is your date of birth?	Are you 18 or older?
Name all the social clubs and organizations to which you belong.	Note any trade or professional memberships that may be relevant to this job.
Have you ever been arrested?	Have you ever been convicted of a crime (if the crime is related to the job requirements)?
Has your family had any health problems recently?	After the hiring process has been completed and if a job offer is made, you will be required to take a medical exam. Is this acceptable?

value 2, 3, or 5 years later. Does the position provide the opportunity to learn new skills? In short, negotiating salary involves considering the company's long-term investment in you.

RESPONDING TO THE JOB INTERVIEW

Letter confirming the interview

When an employer invites you in for an interview, it is a good idea to confirm the details in writing. Again, this letter should be less than a page long. The following approach is helpful:

* Begin by saying "thank you."
* Confirm the time and place of the interview. Note that you plan to bring your portfolio.
* Reconfirm your interest in the company and thank the employer again.

Figure 9.2 is a sample letter confirming an interview.

Followup thank-you letter after the interview

Within 24 hours after the interview, write the person or people who interviewed you a short (no more than one page) thank-you letter. The following strategy will help you write a successful letter:

* Say "thank you" in the first paragraph, and note the date and location of the interview.
* Refer to several specifics discussed during the interview.

Date

Ms. Kris Fowler
Manager, Web Development
Web Design Associates
311 Center Street
San Francisco, CA 94102

Dear Ms. Fowler:

Thank you for inviting me for an interview for the Web designer position.

The date and time you mentioned in our phone conversation work well with my schedule. I will arrive at Web Associates at 10:00 on June 4 and will come directly to your office, as we discussed. Because you requested to see some of the Web pages I have created, I will bring both my electronic and paper portfolios so that you can review my work.

I look forward to meeting you on June 4 and to learning more about Web Design Associates.

Sincerely,

Deborah Jensen

FIGURE 9.2
Sample Letter Confirming the Interview

* Mention or, if you feel that it is appropriate, send your electronic portfolio so that the interviewer can take a second look.
* Finish by sounding positive and upbeat. Provide any contact information that may be needed.

Figure 9.3 is a sample thank-you letter.

Letter declining a job offer

There may come a time when you decide to turn down a job offer (if you are looking for a job right now, we know that this situation seems remote). Keep the letter short, no more than half a page. The following approach is useful:

* Begin on a positive or neutral note.
* State tactfully your reasons for turning down the position.
* Express good will by saying something positive about the company.

Figure 9.4 shows a sample letter declining a job interview.

Letter accepting a job offer

This is the letter you've been waiting to write, and you will enjoy writing it. Again, make it short (no more than one page) and sound enthusiastic. The following tips will help you write an enthusiastic letter of acceptance:

Date

Mr. Harold Williams
Human Resources Manager
Document Design Associates
4500 Main Street
Trenton, New Jersey 08601

Dear Mr. Williams:

Thank you for taking the time to interview me for the entry-level writing position. I was particularly impressed with the state-of-the-art usability lab that Document Design Associates has.

As you requested, I am enclosing two additional copies of writing samples from my portfolio. I am also enclosing a CD of my electronic portfolio should you wish to review any of my projects again.

I am looking forward to hearing from you within the next 2 weeks, as we discussed. If you need any additional information, please contact me at jgleason@comcast.net.

Sincerely,

James Gleason

Enc: 2 writing samples, portfolio CD

FIGURE 9.3
Sample Thank-You Letter

Date

Ms. Leander Morris
Manager, Documentation Group
EZ to Use Software
1733 Union Avenue
Ames, Iowa 50010

Dear Ms. Morris:

Thank you for meeting with me on May 23 to discuss the full-time position as a software documentation writer. I enjoyed meeting you and the other writers on the documentation team.

Since we met, I have been offered another full-time writing position that is a better match with my skills and qualifications. I have decided to accept this other position.

Again, thank you for the opportunity to interview for the writing position with EZ to Use Software.

Sincerely,

Amy Brown

FIGURE 9.4
Sample Letter Declining a Job Offer

* Begin with a clear statement of acceptance.
* Repeat the starting date and add any points that need to be confirmed.
* Restate the good news and end on a positive note.

Figure 9.5 is a sample acceptance letter.

Date

Ms. Kris Fowler
Manager, Web Development
Web Design Associates
311 Center Street
San Francisco, CA 94102

Dear Ms. Fowler:

I am delighted to accept your offer to join Web Design Associates as a Web designer.

As you mentioned in your offer letter, I will plan to begin on June 1 and will report to Gene Stewart at 8:30 for my orientation.

I look forward to a long and rewarding career with Web Design Associates.

Sincerely,

Deborah Jensen

FIGURE 9.5
Sample Letter Accepting a Job Offer

SUMMARY

Chapter 9 offers tips on how to prepare for a successful job interview and how to use your portfolio effectively during the interview. Preparing thoroughly for an interview involves researching each company and learning as much as you can ahead of time about its products, services, organizational philosophy, and interviewing procedure to help you determine if you are a good fit. Anticipating the types of questions that might be asked and preparing answers will help you succeed at the interview. Of equal or even greater importance is making sure that you use your portfolio fully to your advantage. Chapter 9 includes tips on how to tailor your portfolio for each interview and presents guidelines on presenting it to illustrate your best projects and skills.

ASSIGNMENTS

Assignment 1: *Internet Search*

Conduct an Internet search for three companies that employ technical or professional communicators. Write a brief profile of each company, noting its products, services, management philosophy, interviewing style, and any other information that would help you prepare for an interview. If your instructor requests it, be prepared to share your findings with your classmates.

Assignment 2: *Preparing Interview Questions*

Select one of the companies you researched in Assignment 1 and prepare a list of questions that you might be asked at the interview. Write down the answers.

Assignment 3: *Fact-Finding Interview*

Conduct a 15- to 20-minute information-gathering interview with a professional communicator. Your objective is to write both a summary and an evaluation of the interview noting the tasks performed, the tools used, and the communication skills needed to succeed as a professional communicator in the twenty-first century. Before conducting the interview, write a profile of the communicator's company. You should be able to get much of this information from the company's Web site.

You will turn in three documents: (1) a one-page company profile, (2) the list of questions you used to structure the interview, and (3) a two-page summary of the information covered, evaluating what you learned from the interview.

Assignment 4: *Conduct a Mock Interview*

Conduct a mock employment interview with another classmate. You and your partner will each role-play the parts of the interviewee and interviewer. Each interview should last approximately 20 minutes. Include a list of questions from which your partner can choose.

To prepare your partner for the interview, give him or her a detailed job description, resume, letter of application, and selected portfolio pieces. During the interview, practice talking about your portfolio and tailoring your discussion to that particular company. You might also go through an oral presentation of your electronic portfolio to familiarize yourself with its navigation and flow in an interview setting. If you are a professional communicator rather than a student, your mock interview can be run with a mentor or an associate.

REFERENCES

Bolles, Richard N. *What Color Is Your Parachute? A Practical Manual for Job-Hunters and Career-Changers,* 33rd ed. Berkeley, CA: Ten Speed Press, 2004.

"How to Behave in a Behavior-Based Interview." *In Planning Job Choices 2000,* 43rd ed. Bethlehem, PA: National Association of Colleges and Employers, 2000.

Kaffer, Sharon. "Behavioral Interviewing: Ambush or Insight?" *Interviewing at a Job Expo* (2002). http://www. nwcareerexpo. com/interviewing.htm (accessed December 2, 2004).

Molisani, Jack. "Portfolios: Tools for Acing the Interview." *Intercom* 50, no. 8 (September–October 2003): 20–22.

Society for Technical Communication. *2005 Technical Communicator Salary Survey.* Washington, DC: Society for Technical Communication, 2005. http://www.stc.org/salarySurvey.asp (accessed August 3, 2005).

Washington, Tom. *Interview Power.* Bellevue, WA: Mount Vernon Press, 2004.

Conclusion

The portfolio I have built will be a part of my regular life. Actually, it already is. I've shown it to relatives who never really understood what I was doing in school (half of them thought I was a journalism major, and the other half thought I was learning how to write instructions that come with a VCR). I've shown it to friends who thought technical communication was an easy degree (they don't think that so much anymore). I've shown it to people I hope will hire me. I've shown it to anyone who will sit still long enough to look at it. And every couple of days, I sit and look through it myself. Michel

In reflecting on the portfolio-building experience, Michel emphasizes her pride in her portfolio and how the portfolio has become a central part of her professional identity. Throughout this book, we have included student success stories, like Michel's, to support the process approach to creating portfolios that we have described in detail. We conclude *Portfolios for Technical and Professional Communicators* with student comments on the lessons they learned in creating and using their portfolios.

Creating a portfolio helped Brian W. decide what career path to follow. Commenting on the portfolio-building process, Brian focuses on how the revising process helped him rethink his career goals. He says, "Many of the works now have a professional tone and look to them. Some of the writings from this class made me take a step back to rethink who I am and where I want to go." Brian W. credits the portfolio-building process with helping him decide to start his own multimedia business. For Miranda, creating a portfolio helped her make the important transition from thinking of herself as a student to thinking of herself as a professional. She remarks, "Going through the guided process of creating my professional portfolio really set me on the path of thinking of myself and presenting myself as a professional."

For Wylie, creating a portfolio was important because it gave him a better understanding of the careers open to technical and professional communicators:

> This project gave me a greater respect for many different jobs. I acted as a photographer, photograph editor, design team, draftsman, column writer, feature writer, copyeditor, general editor, page layout specialist, printer, publisher, and more.

He concludes that creating a portfolio and using it as a marketing tool will help him get "an entry-level position in any of the aforementioned careers." For Judith, the process of creating a portfolio had intrinsic value. She remarks:

> Overall I really enjoyed the process of creating a portfolio. I learned a lot, but just loved the process of going over my writing and creating a totally professional object that I can show to potential employers.

Nanette credits her mentor for helping her enrich the content of her portfolio by giving her "good ideas about things to include in my portfolio that hadn't occurred to me before." Tom, on the other hand, focuses on the technology skills he learned while creating his electronic portfolio. He remarks that he now knows how to format his graphics so that they download quickly and still retain sharp resolution. For Norma, creating her portfolio was a transforming event. She describes the experience as a metamorphosis and observes how her portfolio records her "development from a computer science major into a technical communication professional."

Students also had advice for those who might take on the task of creating a portfolio. "Save everything you do because you will need it," cautions Wilda. Wylie's advice focuses on audience analysis as he stresses the importance of writing "each assignment with your future employer in mind." Joy advises those beginning the portfolio-building process not to dismiss work created outside the classroom, particularly "volunteer work, contract jobs, and internships."

Students also commented on how their portfolios contributed to a successful job search. Miranda says that she marketed herself by sending potential employers a link to her online portfolio. While she felt hesitant about showing all her work without being present, she notes that the exposure resulted in "those meaningful call backs—what time would you be available for an interview?" After being hired, Miranda was pleasantly surprised by the compliments she received from coworkers who had seen her online portfolio. Brian D. also used his portfolio to complete a successful job search with a small environmental company. He mentions that he brought his paper portfolio to an interview with extra samples that pertained specifically to the job requirements. Showing his work helped get him the job offer. For Sarah, using her portfolio at an interview helped her better explain her skills and accomplishments because "it was so much easier showing an interviewer what I could do than trying to explain what I could do."

All of these student comments emphasize the importance of having a professional portfolio that can help you market your skills and abilities as a professional communicator. Following the process approach discussed throughout this book will help you create a successful professional portfolio. You will also feel more confident about using the portfolio during your job search.

Finally, updating your portfolio after beginning your professional career is also very important, as Sarah notes in the following comment. After using her portfolio successfully to find a full-time position, Sarah realized how important it was to keep it current by adding new material that highlighted her professional development:

> I update my portfolio on a regular basis. I gather up the invitations, business cards, and meeting presentations and materials that I've designed and make new pages for them in my portfolio. I try to keep it as updated as possible so that I can grab it any time and back up my design skills.

Your portfolio is an evolving document, changing as you change, reflecting your professional growth and development.

References and Additional Resources

Portfolios in General

Allen, Jo. "Assessment Methods for Business Communication: Tests, Portfolios, and Surveys." *Bulletin of the Association for Business Communication* 61, no. 3 (1998): 64–66.

Ause, Cheryl Evans, and Gerilee Nicastro. "Establishing Sound Portfolio Practice: Reflections on Faculty Development." In *Situating Portfolios Four Perspectives*, eds. Kathleen Blake Yancy and Irwin Weiser. Logan: Utah State University Press, 1997.

Barnum, Carol. *Usability Testing and Research*. New York: Longman, 2002.

Belanoff, Pat, and Marcia Dickson, eds. *Portfolios: Process and Product*. Portsmouth, NH: Boynton/Cook-Heinemann, 1991.

Bishop, Wendy. "Designing a Portfolio Evaluation System." In *Teaching Lives: Essays and Stories*. Logan: Utah State University Press, 1991.

_____. *Acts of Revision: A Guide for Writers*. Portsmouth, NH: Boynton/Cook, 2004.

Bitzer, Lloyd. "The Rhetorical Situation." *Philosophy and Rhetoric* 1 (1968): 1–14.

Burch, C. Beth. "Inside the Portfolio Experience: The Student's Perspective." *English Education* 32, no. 1 (1999): 34–49.

Campbell, Dorothy M., Pamela Bondi Cignetti, Beverly J. Melenyzer, Diane H. Nettles, and Richard M. Wyman. *How to Develop a Professional Portfolio*, 2nd ed. Boston: Allyn & Bacon, 2001.

Campbell, Nittaya. "Getting Rid of the Yawn Factor: Using a Portfolio Assignment to Motivate Students in a Professional Writing Class." *Business Communication Quarterly* 65, no. 3 (2002): 42–55.

Carlson, Randal D. "Portfolio Assessment of Instructional Technology." *Journal of Educational Technology Systems* 27, no. 1 (1998): 81–92.

Cooper, Winfield, and B. J. Brown. "Using Portfolios to Empower Student Writers." *English Journal* 81 (1992): 40–45.

Danielson, Charlotte, and Leslie Abrutyn. *An Introduction to Using Portfolios in the Classroom*. Alexandria, VA: Alexandria Association for Supervision and Curriculum Development, 1997.

Davis, Robert Leigh. "The Lunar Light of Student Writing: Portfolios and Literary Theory." In *Situating Portfolios: Four Perspectives*, eds. Kathleen Blake Yancy and Irwin Weiser. Logan: Utah State University Press, 1997.

Dillon, W. Tracy. "Corporate Advisory Boards, Portfolio Assessment, and Business and Technical Writing Program Development." *Business Communication Quarterly* 60, no. 1 (1997): 41–58.

Elbow, Peter and Pat Belanoff. "Reflections on an Explosion; Portfolios in the '90s and Beyond." In *Situating Portfolios: Four Perspectives*, eds. Kathleen Blake Young and Irwin Weiser. Logan: Utah State University Press, 1997.

Farr, Roger. *Portfolios: Assessment in the Language Arts*. New York: HBJ College Publishers, 1991.

Fishman, Stephen. *The Public Domain: How to Find, Copyright-Free Writings, Music, Art & More*. Berkeley, CA: Nolo Press, 2000.

Foster, Andrea. "Library Groups Say Sweeping State Copyright Laws Could Stifle Teaching and Research." *Chronicle of Higher Education* 49, no. 32 (April 1, 2003): 1–5.

Frechette, Julie D. *Developing Media Literacy in Cyberspace: Pedagogy and Critical Learning for the Twenty-First-Century Classroom*. Westport, CT: Greenwood, 2002.

Gomez, Emily. "Assessment Portfolios: Including English Language Learners in Large-Scale Assessments." *ERIC Digest* (December 2000). http://www.cal.org/resources/digest/0010assessment.html (accessed June 2, 2004).

Herbeck, Dale A., and Christopher D. Hunter. "Intellectual Property in Cyberspace: The Use of Protected Images on the World Wide Web." *Communication Research Reports* 15 (1998): 57–63.

Herbert, E. "Portfolios Invite Reflection—from Students and Staff." *Educational Leadership* 49, no. 8 (1992): 58–60.

Herbert, E., and L. Schultz. "The Power of Portfolios." *Educational Leadership* 53, no. 7 (1996): 70–71.

Hoger, Elizabeth A. "A Portfolio Assignment for Analyzing Business Communications." *Business Communication Quarterly* 61, no. 3 (1998): 64–66.

Huot, Brian, and Michael M. Williamson. "Rethinking Portfolios for Evaluating Writing: Issues of Assessment and Power." In *Situating Portfolios: Four Perspectives*, eds. Kathleen Blake Yancy and Irwin Weiser. Logan: Utah State University Press, 1997.

Jameson, Daphne. "The Ethics of Plagiarism: How Genre Affects Writers' Use of Source Materials." *Bulletin of the Association for Business Communication* 56, no. 2 (1993): 18–28.

Kastman, Lee-Ann M., and Laura J. Gurak. "Conducting Technical Communication Research Via the Internet: Guidelines for Privacy, Permissions, and Ownership in Educational Research." *Technical Communication* 46 (1999): 460–469.

Keirsey-Temperament Sorter. www.keirsey.com (accessed August 15, 2005).

Lakoff, George, and Mark Johnson. *Metaphors We Live By.* Chicago: University of Chicago Press, 1980.

Linton, Harold. *Portfolio Design*, 3rd ed. New York: W.W. Norton, 2003.

Luescher, Andreas. "The Professional Portfolio as Heuristic Methodology." *Journal of Technical Writing and Communication* 32 (2002): 353–366.

Martin, Debra Bayles. *The Portfolio Planner: Making Professional Portfolios Work for You.* Upper Saddle River, NJ: Prentice Hall, 1999.

McDaniel, Barbara, and Roberta Sheng-Taylor. "The Communicator's Portfolio: The Ideal Saleskit." *Technical Communication* 39, no. 2 (May 1992): 215–218.

Molisani, Jack. "Portfolios: Tools for Acing the Interview." *Intercom* 50, no. 8 (September–October 2003): 20–22.

Murray, Mary Ellen. "The Semester Portfolio." *Bulletin of the Association for Business Communication* 57, no. 1 (1994): 1–5.

Myers-Briggs Personality Type Indicator. *The Myers-Briggs Foundation.* www.myersbriggs.org (accessed August 15, 2005).

Plumb, Carolyn, and Cathie Scott. "Outcomes Assessment of Engineering Writing at the University of Washington." *Journal of Engineering* 91 (2002): 333–338.

Powell, Karen Sterkel, and Jackie L. Jankovich. "Student Portfolios: A Tool to Enhance the Traditional Job Search." *Business Communication Quarterly* 61, no. 4 (1998): 72–82.

Purves, Alan J., Joseph A. Quattrini, and Christine I. Sullivan. *Creating the Writing Portfolio: A Guide for Students.* Lincolnwood, IL: NTC Publishing, 1995.

Satterthwaite, Frank, and Gary D'Orsi. *The Career Portfolio Workbook.* New York: McGraw-Hill, 2003.

Scott, Cathie, and Carolyn Plumb. "Using Portfolios to Evaluate Service Courses as Part of an Engineering Writing Program." *Technical Communication Quarterly* 8 (1999): 337–350.

Scott, Julie S. "Portfolios for Technical Communicators: Worth the Work." *Intercom* 47, no. 2 (February 2000): 26–28.

Seldin, Peter. "The Teaching Portfolio." *ASEE Prism* 4, no. 9 (1995): 19–22.

Smith, Mary Ann. "Behind the Scenes: Portfolios in a Classroom Learning Community." In *Situating Portfolios: Four Perspectives*, eds. Kathleen Blake Yancy and Irwin Weiser. Logan: Utah State University Press, 1997.

Sormunen, Carolee. "Portfolios: An Assessment Tool for School-to-Work Transition." *Business Education Forum* 48 (April 1994): 8–10.

Straub, Richard, and Ronald F. Lunsford. *Twelve Readers Readings: Responding to College Student Writing.* Cresskill NJ: Hampton Press, 1995.

Twain, Mark. *A Connecticut Yankee in King Arthur's Court.* New York: Signet Classics, 2004.

University of Washington. "PEP's Original Goals." October 2004. http://www.engr.washington.edu/abet/PEP%20Presentation/tsld002.htm (accessed May 23, 2005).

Williams, Anna Graf and Karen J. Hall. *Creating Your Career Portfolio: At a Glance Guide for Students*, 2nd ed. Upper Saddle River, NJ: Prentice Hall, 2001.

Williams, Julia M. "Technical Communication, Engineering, and ABET's Engineering Criteria: 2000 What Lies Ahead?" *Technical Communication* 49 (2002): 89–95.

Yancy, Kathleen Blake, and Irwin Weiser, eds. *Situating Portfolios: Four Perspectives.* Logan: Utah State University Press, 1997.

Career-Related Sources

Arn, Joseph V., Rebecca Gatlin, and William Kordsmeier. "Multimedia Copyright Laws and Guidelines: Take the Test." *Business Communication Quarterly* 61, no. 4 (1998): 32–39.

Bailie, Rahel Anne. "Using a Résumé to Showcase Your Talents." *Intercom* 50, no. 7 (September–October 2003): 15–19.

Blackburn-Brockman, Elizabeth, and Kelly Belanger. "One Page or Two?: A National Study of CPA Recruiters' Preferences for Résumé Length." *Journal of Business Communication* 38 (2001): 29–45.

Bloch, Janel M. "Online Job Searching: Clicking Your Way to Employment." *Intercom* 50, no. 8 (September–October 2003): 11–14.

Block, Barbara M. "Finding That First Job." *Intercom* 48, no. 5 (May 2001): 22–24.

Board of Regents of the University System of Georgia Office of Legal Affairs. *Regents Guide to Understanding Copyright and Educational Fair Use.* http://www.usg.edu/admin/legal/copyright (accessed February 10, 2003).

Bolles, Richard N. *Job Hunting on the Internet*, 2nd ed. Berkeley, CA: Ten Speed Press, 1999.

————. *What Color Is Your Parachute? A Practical Manual for Job-Hunters and Career-Changers*, 33rd ed. Berkeley, CA: Ten Speed Press, 2004.

Bosley, Deborah S. "Collaborative Partnerships: Academia and Industry Working Together." *Technical Communication* 42 (1995): 611–19.

Botan, Carl. "Ethics in Strategic Communication Campaigns: The Case for a New Approach to Public Relations." *Journal of Business Communication* 34 (1997): 160–70.

Caher, John M. "Technical Documentation and Legal Liability." *Journal of Technical Writing and Communication* 25, no. 1 (1995): 5–10.

Campbell, Charles P. "Ethos: Character and Ethics in Technical Writing." *IEEE Transactions on Professional Communication* 38 (1995): 132–38.

Christians, Clifford, and Michael Trabers, eds. *Communication Ethics and Universal Values*. Thousand Oaks, CA: Sage, 1997.

Clark, Thomas. "Teaching Students How to Write to Avoid Legal Liability." *Business Communication Quarterly* 60, no. 3 (1997): 71–77.

Conlin, Michelle. "And Now, the Just-in-Time-Employee." *Business Week*, no. 3696 (August 2000) http://web27. epnet.com/citation.asp (accessed November 19, 2004).

Crawford, Tad, and Kay Murray. *The Writer's Legal Guide: An Author's Guild Reference*, 3rd ed. New York: Allworth Press, 2002.

Crowther, Karmen N. T. "How to Research Companies." In *Planning Job Choices 2000*, 43rd ed. Bethlehem, PA: National Association of Colleges and Employers, 1999.

DeLain, Nancy. "Intellectual Property for Technical Communicators." April 22, 2004. http://www-106.ibm.com/developerworks/rational/Library/4595 (accessed August 26, 2004).

Dombrowski, Paul M. *Ethics in Technical Communication*. Boston: Allyn & Bacon, 2000.

Dragga, Sam. "Ethical Intercultural Technical Communication: Looking through the Lens of Confucian Ethics." *Technical Communication Quarterly* 8 (1999): 365–81.

Faber, Brenton. "Intuitive Ethics: Understanding and Critiquing the Role of Intuition in Ethical Decisions." *Technical Communication Quarterly* 8 (1999): 189–202.

Glick-Smith, Judith L., and Carol Stephenson. "Why Do Contractors and Independent Consultants Need Lawyers?" *Technical Communication* 45 (February 1998): 89–94.

Grice, Roger A. "STC's Communities of Practice—Networking with Our Peers." *Intercom* 51, no. 3 (March 2004): 28–29.

Gurak, Laura J. "Technical Communication, Copyright, and the Shrinking Public Domain." *Computers and Composition* 14 (1997): 329–42.

Halbert, Deborah J. *Intellectual Property in the Information Age: The Politics of Expanding Ownership Rights*. Westport, CT: Quorum, 1999.

Hart, Geoffrey J. S. "Effective Interviewing: Get the Story." *Intercom* 47, no. 1 (January 2000): 24–26.

Hartung, Kris K. "What Are Students Being Taught about the Ethics of Technical Communication?: An Analysis of Ethical Discussions Presented in Four Textbooks." *Journal of Technical Writing and Communication* 28 (1998): 363–83.

Hawthorne, Mark D. "Learning by Doing: Teaching Decision Making through Building a Code of Ethics." *Technical Communication Quarterly* 10 (2001): 341–55.

Helyar, Pamela S., and Gregory M. Doudnikoff. "Walking the Labyrinth of Multimedia Law." *Technical Communication* 50 (2003): 497–504.

Herbeck, Dale A., and Christopher D. Hunter. "Intellectual Property in Cyberspace: The Use of Protected Images on the World Wide Web." *Communication Research Reports* 15 (1998): 57–63.

Herrington, TyAnna K. "Ethics and Graphic Design; A Rhetorical Analysis of the Document Design in the *Report of the Department of the Treasury on the Bureau of Alcohol, Tobacco, and Firearms Investigation of Vernon Wayne Howell also Known as David Koresch*." *IEEE Transactions on Professional Communication* 38 (1995): 151–57.

————. "Who Owns My Work? The State of Work for Hire for Academics in Technical Communication." *Journal of Business and Technical Communication* 13, no. 2 (1999): 125–53.

————. "Work for Hire for Nonacademic Creators." *Journal of Business and Technical Communication* 13, no. 4 (1999): 401–26.

————. *A Legal Primer for the Digital Age*. New York: Pearson Longman, 2003.

Jensen, J. Vernon. *Ethical Issues in the Communication Process*. Mahwah, NJ: Erlbaum, 1997.

Jones, Dan, and Karen Lane. *Technical Communication: Strategies for College and the Workplace*. Boston: Allyn & Bacon/Longman, 2001.

Kaffer, Sharon. "Behavioral Interviewing: Ambush or Insight?" *Interviewing at a Job Expo* (2002). http://www.nwcareerexpo.com/interviewing.htm (accessed December 2, 2004).

Kant, Immanuel. *Groundwork of the Metaphysics of Morals*. Translated by H. J. Paton. New York: Harper & Row, 1964.

Kastman, Lee-Ann M., and Laura J. Gurak. "Conducting Technical Communication Research via the Internet:

Guidelines for Privacy, Permissions, and Ownership in Educational Research." *Technical Communication* 46 (1999): 460–69.

Keefer, Christine A. "From Not Working—To Networking." *Intercom* 51, no. 3 (March 2004): 4–7.

Kienzler, Donna. "Ethics, Critical Thinking, and Professional Communication Pedagogy." *Technical Communication Quarterly* 10 (2001): 319–39.

Kimeldorf, Martin. *Creating Portfolios for Success in School, Work, and Life.* Minneapolis: Free Spirit, 1994.

Krause, Tim. "Preparing an Online Resume." *Business Communication Quarterly* 60, no. 1 (1997): 159–61.

Lambe, Jennifer L. "Techniques for Successful SME Interviews." *Intercom* 47, no. 3 (March 2000): 30–32.

Lannon, John M. *Technical Communication*, 9th ed. New York: Longman, 2003.

La Plante, Alice. "Liability in the Information Age." *InfoWorld* 8, no. 33 (August 18, 1986): 37–38.

LaVie, Donald S., Jr., "Internet Technology and Intellectual Property." *Intercom* 47, no. 1 (January 2000): 20–23.

Markel, Mike. "An Ethical Imperative for Technical Communicators." *IEEE Transactions on Professional Communication* 36 (1993): 81–86.

———. "Ethics and Technical Communication: A Case for Foundational Approaches." *IEEE Transactions on Professional Communication* 40 (1997): 284–98.

———. "Deep Linking: An Ethical and Legal Analysis." *IEEE Transactions on Professional Communication* 45 (2002): 77–83.

McQuaide, Judith, Gaea Leinhardt, and Catherine Stainton. "Ethical Reasoning: Real and Simulated." *Journal of Educational Computing Research* 21 (1999): 433–74.

Mill, John Stuart. "Utilitarianism." In *The Utilitarians.* Garden City, NY: Dolphin Books, 1998.

O'Rourke, James S. "The Ethics of Resumés and Recommendations: When Do Filler and Fluff Become Deception and Lies?" *Business Communication Quarterly* 58, no. 4 (1995): 54–56.

Parietz, Beth. "Ethics Instruction: An Undergraduate Essential." *ASEE Prism* 5, no. 2 (1995): 20–25.

Paul, Jim, and Christy A. Strbiak. "The Ethics of Strategic Ambiguity." *Journal of Business Communication* 34 (1997): 147–48.

Pfeiffer, William Sanborn. *Technical Writing: A Practical Approach*, 5th ed. Upper Saddle River, NJ: Prentice Hall, 2003.

Ralston, Steven M. "Teaching Interviewees Employment Interviewing Skills: A Test of Two Alternatives." *Journal of Business and Technical Communication* 9 (1995): 362–69.

The Republic of Plato. Translated by F. M. Cranford. New York: Oxford University Press, 1962.

Robart, Kay, and K. C. Francis "Mentoring in a Business Environment." *Intercom* 48, no. 8 (September–October 2001): 26–28.

Roever, Carol. "Preparing the Scannable Resume." *Business Communication Quarterly* 60, no. 1 (1997): 156–59.

Ryesky, Kenneth H. "The Effects of Print and Other Text Media Developments upon the Law in America." *Journal of Technical Writing and Communication* 29 (1999): 3–30.

Sageev, Pneena, and Carol J. Romanowsky. "A Message from Recent Engineering Graduates in the Workplace: Results of a Survey on Technical Communication Skills." *Journal of Engineering Education* 90 (October 2001): 685–93.

Samuelson, Pamela. "Toward a New Politics of Intellectual Property." *Communications of the ACM* 44 (2001): 98–99.

Sanders, Scott P., and M. Jimmie Killingsworth. "Ethics in Professional Communication." *IEEE Transactions on Professional Communication* 38 (1995): 129–31.

Savage, Gerald J., and Dale L. Sullivan. *Writing: A Professional Life.* Boston: Allyn & Bacon, 2001.

Scholl, Christopher J. "Technology and Communication Ethics: An Evaluative Framework." *Technical Communication Quarterly* 4 (1995): 157–64.

Scott, J. Blake. "Sophistic Ethics in the Technical Writing Classroom: Teaching Nomos, Deliberation, and Action." *Technical Communication Quarterly* 4 (1995): 187–99.

Searles, George J. *Workplace Communications: The Basics*, 2nd ed. Boston: Allyn & Bacon/Longman, 2003.

Simpson, Carol. *Copyright for Schools: A Practical Guide*, 4th ed. Worthington, OH: Linwood Publishers, 2005.

Sims, Brenda. "Linking Ethics and Language in the Technical Communication Classroom." *Technical Communication Quarterly* 2 (1993): 285–99.

Sims, Ronald R. *Teaching Business Ethics for Effective Learning.* Westport, CT: Greenwood, 2002.

Smith, Howard T., and Henrietta Nickels Shirk. "The Perils of Defective Documentation." *Journal of Business and Technical Communication* 10, no. 2 (1996): 187–202.

Society for Technical Communication. *Ethical Principles for Technical Communicators.* Washington, DC: Society for Technical Communication, 2005. http://www.stc.org/policy_statements_ethicalPrinciples.asp (accessed January 24, 2005).

Society for Technical Communication. 2005 *Technical Communicator Salary Survey* (2005). http://www.stc.org/salarySurvey.asp (accessed August 3, 2005).

Stim, Richard. *Getting Permission: How to License and Clear Copyrighted Materials Online and Off*, 2nd ed. Berkeley, CA: Nolo Press, 2004.

Studevant, Mirhonda. "Consulting Pros and Cons." *Intercom* 45, no. 9 (November 1998): 20–22.

Tucker, Robert B. "Recruiting Strategies That Get Results." *Intercom* 48, no. 7 (July–August 2001): 12–13.

Twain, Mark. *Mark Twain's Notebook*, ed. Albert B. Paine. New York: Cooper Square, 1972.

U.S. Bureau of Labor. *U.S. Bureau of Labor Statistics Occupation Outlook Handbook (2004–5)*. http://www.bls.gov/oco/ocos089.htm (accessed July 6, 2004).

U.S. Copyright Office. "The Digital Millennium Copyright Act of 1998." Washington, DC: Library of Congress Copyright Office, 1998, pp. 1–18. http://www.copyright.gov/legislation/dmca.pdf (accessed April 6, 2005).

U.S. Copyright Office. "Highlights of Copyright Amendments Contained in the Uruguay Round Agreements Act (URAA)" (Circular 38b). Washington, DC: Library of Congress Copyright Office, 2003.

U.S. Copyright Office. "International Copyright Relations of the United States" (Circular 38a). Washington, DC: Library of Congress Copyright Office, 2003.

U.S. Copyright Office. "Work Made for Hire Under the 1976 Copyright Act" (Circular 9). Washington, DC: Library of Congress, 2003.

U.S. Copyright Office. "Copyright Law of the United States: Section 107—Limitations on Exclusive Rights: Fair Use" (Circular 92). Washington, DC: Library of Congress, 2004.

U.S. Department of Labor, Employment and Training Administration. *Tips for Finding the Right Job* (1996). http://www.doleta.gov/uses/tips4jobs.pdf (accessed July 6, 2004).

U.S. Department of Labor, Office of Disability Employment Policy. "Writing and Formatting a Scannable Resume" (February 2005). http://www.dol.gov/odep/pubs/ek99/resume.htm (accessed February 5, 2005).

U.S. Department of Labor Employment and Training Administration. "Tips for Finding the Right Job." (1996). http://www.doleta.gov/uses/tip4jobs.pdf (accessed May 25, 2004).

Van Wicklin, Janet. *The Tech Writer's Survival Guide*. New York: Checkmark Books, 2001.

Washington, Tom. *Interview Power: Selling Yourself Face to Face*. Bellevue, WA: Mount Vernon Press, 2004.

Willis, Jim, Albert Adelowo Okunade, and William James Willis. *Reporting on Risks: The Practice and Ethics of Health and Safety Communication*. Westport, CT: Praeger, 1997.

Electronic Portfolios and Graphic Design

Arnheim, Rudolf. *Visual Thinking*. Berkeley: University of California Press, 1969.

Barrett, Helen C. "Create Your Own Electronic Portfolio: Using Off-the-Shelf Software to Showcase Your Own or Student Work." *Learning and Leading with Technology* 27, no. 7 (2000): 14–21.

———. "Electronic Portfolios = Multimedia Development + Portfolio Development: The Electronic Portfolio Development Process," 2000. http://transition.alaska.edu/www/portfolios/aahe2000.html (accessed March 18, 2005).

Berryman, Gregg. *Notes on Graphic Design and Visual Communication*. Crisp Publications, 1990.

Bertoline, Gary R., Craig Miller, and Leonard O. Nasman. *Technical Graphics Communication*. Chicago: Irwin: 1995.

Blair, Kristine L., and Pamela Takayoshi. "Reflections in Reading and Evaluating Electronic Portfolios." In *Situating Portfolios: Four Perspectives*, eds. Kathleen Blake Yancey and Irwin Weiser. Logan: Utah State University Press, 1997.

Bryan, John. "Seven Types of Distortion: A Taxonomy of Manipulative Techniques Used in Charts." *Journal of Technical Writing and Communication* 25 (1995): 127–80.

Campbell, Jo. "Electronic Portfolios: A Five-Year History." *Computers and Composition* 13, no. 2 (1996): 185–94.

Dikel, Margaret F., ed. "Resumes and Cover Letters: Prepare Your Resume for E-Mailing or Posting on the Internet." *The Riley Guide*. http://www.rileyguide.com/letters.html (accessed April 25, 2005).

Dragga, Sam, and Dan Voss. "Cruel Pies: The Inhumanity of Technical Illustrations." *Technical Communication* 48 (2001): 325–74.

Dubinsky, Jim. "Creating New Views on Learning: e-Portfolios." *Business Communication Quarterly* 66, no. 4 (2003): 96–102.

Gonzalez de Cosio, Maria, and Sharon Helmer Poggendohl, eds. *Visual Rhetoric*. Special Issue of *Visible Language* 32, no. 3 (1998).

Gurak, Laura J. "Toward Consistency in Visual Information: Standardized Icons Based on Task." *Technical Communication* 50 (2003): 492–96.

Hansen, Katharine. "Your E-Resume's File Format Aligns with Its Delivery Method." http://www.quintcareers.com/e-resume_format.html (accessed April 25, 2005).

———. "The Top 10 Things You Need to Know about E-Resumes and Posting Your Resume Online." http://www.quintcareers.com/e-resumes.html (accessed April 25, 2005).

Horton, William. "The Almost Universal Language: Graphics for International Documents." *Technical Communication* 40 (1993): 682–93.

Howath, Clara. "Electronic Resume." http://www.damngood.com/jobseekers/electronic.html (accessed May 10, 2005).

Howard, Rebecca Moore. "Memoranda to Myself: Maxims for the Online Portfolio." *Computers and Composition* 13, no. 2 (1996): 155–67.

Juillet, Christopher. "Protect Your Web Site from Legal Land Mines." *Intercom* 51, no. 9 (November 2004): 6–7.

Kendus, Steven M. "Developing a Web-Based Portfolio." *Intercom* 49, no. 9 (November 2002): 4–7.

Kienzier, Donna S. "Visual Ethics." *Journal of Business Communication* 34 (1997): 171–87.

Kimball, Miles A. *The Web Portfolio Guide.* New York: Longman, 2003.

Knadler, Stephen. "E-Racing Difference in E-Space: Black Female Subjectivity and the Web-Based Portfolio." *Computers and Composition* 18 (2001): 235–55.

Moore, Patrick, and Chad Fitz. "Using Gestalt Theory to Teach Document Design and Graphics." *Technical Communication Quarterly* 2 (1993): 389–410.

Powley, William. "Technical and Scientific Illustrations: From Pen to Computer." *STC Conference Proceedings 2005.* http://www.stc.org/Confproceed/1995/PDFs/PG349352 (STC members only; accessed July 26, 2004).

Purves, Alan C. "Electronic Portfolios." *Computers and Composition* 13, no. 2 (1996): 135–46.

Ryan, Colin. *Exploring Perception.* Pacific Grove, CA: Brooks/Cole and Nelson ITP, 1997.

Sanchez, Mario. "Web-Safe Fonts for Your Site." http://www.accordmarketing.com/tid/archive/websafefonts.html (accessed October 31, 2005).

Search, Patricia. "Computer Graphics: Changing the Language of Visual Communication." *Technical Communication* 40 (1993): 629–37.

Shelton, S.M. Special Issue: Visual Communication. *Technical Communication* 40 (1993): 617–18.

Smith, Rebecca. *Electronic Resumes and Online Networking*, 2nd ed. Franklin Lakes, NJ: Career Press, 2000.

———. "eResumes 101: Choosing Your Best Electronic Resume Format." http://www.eresumes.com/eresumes_read.html (accessed April 25, 2005).

Sullivan, Patricia. "Practicing Safe Visual Rhetoric on the World Wide Web." *Computers and Composition* 18 (2001): 103–21.

Villano, Matt. "Hi-Octane Assessment: The Electronic Portfolio Powers-up Student, Educator, and Lifelong Assessment." *Campus Technology* (September 2005): 45–48.

Wall, Beverly C., and Robert F. Peltier. "'Going Public' with Electronic Portfolios: Audience, Community, and the Terms of Student Ownership." *Computers and Composition* 13, no. 2 (1996): 207–17.

Watkins, Steve. "World Wide Web Authoring in the Portfolio-Assessed, (Inter) Networked Composition Course." *Computers and Composition* 13, no. 2 (1996): 219–30.

Whitcomb, Susan B., and Pat Kendall. *E-Resumes: Everything You Need to Know About Using Electronic Resumes to Tap into Today's Hot Job Market.* New York: McGraw-Hill, 2001.

White, Jan V. "Color: The Newest Tool for Technical Communicators." *Technical Communication* 50 (2003): 485–91.

Williams, Robin, and John Tollett. *The Non-Designer's Web Book*, 2nd ed. Berkeley, CA: Peachpit Press, 2000.

Williams, Robin. *The Non-Designer's Design Book*, 2nd ed. Berkeley, CA: Peachpit Press, 2003.

Yancy, Kathleen Blake. "Portfolio, Electronic, and the Links Between." *Computers and Composition* 13, no. 2 (1996): 129–33.

———. "The Electronic Portfolio: Shifting Paradigms." *Computers and Composition* 13, no. 2 (1996): 259–62.

Index